Synthesis Lectures on Chemical Engineering and Biochemical Engineering

This series publishes short books on all aspects of chemical engineering, covering the analysis or design of chemical processes to effectively convert materials into more useful materials or energy. The books will focus on fundamental aspects necessary for chemical engineering design including chemistry, math, physics, and sometimes biology to improve the quality of life by inventing, optimizing, and economizing new technologies and products.

Wujie Zhang

Biopolymer Engineering

Springer

Wujie Zhang
Chemical and Biomolecular Engineering
Program, Department of Physics
and Chemistry
Milwaukee School of Engineering
Milwaukee, WI, USA

ISSN 2327-6738   ISSN 2327-6746 (electronic)
Synthesis Lectures on Chemical Engineering and Biochemical Engineering
ISBN 978-3-031-98867-7   ISBN 978-3-031-98868-4 (eBook)
https://doi.org/10.1007/978-3-031-98868-4

This Springer imprint is published by the registered company Springer Nature Switzerland AG
The registered company address is: Gewerbestrasse 11, 6330 Cham, Switzerland

Competing Interests The author has no competing interests to declare that are relevant to the content of this manuscript.

Contents

About the Author

Wujie Zhang is a Professor of Chemical and Biomolecular Engineering at the Milwaukee School of Engineering (MSOE). He received a Ph.D. degree in biomedical engineering from the University of South Carolina and M.S. and B.S. degrees in food science and engineering from the University of Shanghai for Science and Technology. Dr. Zhang's research and scholarly work cover a broad range of topics, including biomaterials, tissue engineering, drug delivery, and food engineering. His current research on biopolymers focuses on designing bioinks for bioprinting and developing innovative drug delivery systems. Dr. Zhang has published over 45 peer-reviewed journal articles in leading journals such as ACS Nano, Acta Biomaterialia, and Polymers. He is also the author of the books *Nanotechnology for Bioengineers* and *Fundamentals and Design of Drug Delivery Systems: A Textbook with Lab Activities*. With over 12 years of experience teaching the Biopolymer Engineering and Polymer Chemistry courses at MSOE, he has developed numerous lab activities to enhance learning. Dr. Zhang's contributions have been widely recognized: he was named one of the 20 high-achieving researchers and educators under 40 by ASEE's Prism magazine (2018), honored as STEM Forward's Young Engineer of the Year (2022), and received MSOE's Oscar Werwarth Distinguished Teacher Award (2023).

Introduction to Polymers and Biopolymers 1

1.1 Definition of Polymers and Biopolymers

The term "polymer" (from the Greek *polys* meaning "many" and *meros* meaning "part') was first introduced by Jöns Jakob Berzelius (Swedish chemist) in 1833 [1]. In 1922, Hermann Staudinger (German chemist) developed the new concept—macromolecules considering the original meaning of the term polymer did not carry any connotation concerning size. The concept of macromolecules covered both synthetic and natural polymers [1, 2]. According to the International Union of Pure and Applied Chemistry (IUPAC), "polymer" is defined as a substance composed of macromolecules; while "macromolecule/polymer molecule" is defined as a molecule of high relative molecular mass, the structure of which essentially comprises the multiple repetition of units derived, actually or conceptually, from molecules of low relative molecular mass [3]. Although a common practice, the terms polymer and macromolecule are not interchangeable considering a polymer is generally a mixture of macromolecules which can differ in chain length and/or structure [4].

The IUPAC defines biopolymers as macromolecules (including proteins, nucleic acids and polysaccharides) formed by living organisms [5]. There are other definitions which are more inclusive, such as a polymer produced from natural sources either chemically synthesized from a biological material or entirely biosynthesized by living organisms [6]. The Office of Technology Assessment of the U.S. Congress defines biopolymers as polymers that are derived from biological precursors or are produced by using modern biotechnology [7]. In the context of this book, to be comprehensive, biopolymers are defined as polymers that fulfill at least one of the following properties [8]:

W. Zhang, *Biopolymer Engineering*, Synthesis Lectures on Chemical Engineering and Biochemical Engineering, https://doi.org/10.1007/978-3-031-98868-4_1

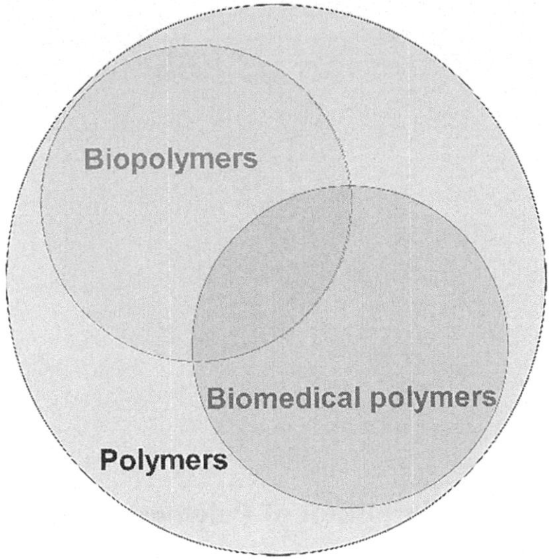

Fig. 1.1 Relationships between polymers, biopolymers, and biomedical polymers

- consist (partly) of bio-based (renewable) raw materials
- are in some way biodegradable.

Based on this definition, biopolymers can be classified into three categories: degradable non-bio-based biopolymers, degradable bio-based biopolymers, and non-degradable bio-based biopolymers [8]. The terms biomedical polymers or polymeric biomaterials have also been widely used but is restricted to biomedical applications, such as drug delivery and tissue engineering [9, 10]. Not all the biomedical polymers are biopolymers based on the biopolymer definitions. For example, polyethylene and polypropylene have been widely used in biomedical fields but they are not biopolymers (non-degradable and non-bio-based) [11]. The relationships between polymers, biopolymers, and biomedical polymers are shown in Fig. 1.1.

1.2 History of Polymers and Biopolymers

By 1600 B.C., ancient Mesoamerican peoples harvested latex from *Casilla elastica* to produce rubber balls and other objects. Polyisoprene is the primary composition of natural rubber [12]. Charles Goodyear discovered the vulcanization of rubber (patented the process in 1844) which revolutionized the rubber industry in the mid-1800s [13, 14]. Vulcanization is a process of cross-linking rubber molecules chemically with organic/inorganic substance, such as sulfur, through the action of heat (140–200 °C) and pressure [15]. During the process, elemental sulfur is dissolved and/or dispersed in a rubber

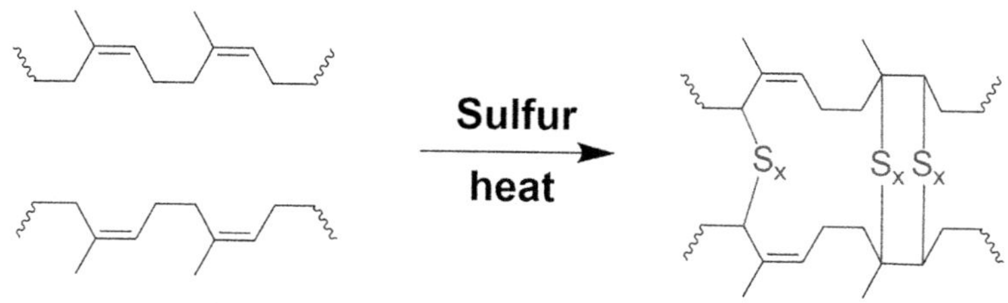

Fig. 1.2 Possible cross-linking mechanisms of natural rubber with sulfur and accelerators

matrix and then subjected to a thermal treatment. This results in the formation of sulfide bonds (-S_X-) connecting adjacent rubber chains, leading to a desired viscoelastic response (Fig. 1.2) [16–19]. In the industrial vulcanization process, vulcanization rate and cross-linking efficiency are improved by using accelerators and such [20].

Cellulose is another important polymer in human history. Cotton is the purest form of cellulose available in nature [21]. During 1830s, both Henri Braconnot and Théophile-Jules Pelouze worked on the nitration of cellulose [22]. Around 1846, Christian Friedrich Schönbein, a German chemist, accidently discovered guncotton (nitrocellulose) (Fig. 1.3). He accidentally spilled nitric acid and sulfuric acid in his wife's kitchen and wiped it up using a cotton apron. When the apron was drying, it spontaneously ignited [23]. Guncotton did not come into use as an ingredient of gunpowder until 1860s. In 1869, John Hyatt, inspired by a New York firm's offer of $ 10,000 for finding a replacement of ivory, invited the first commercially successful plastic—celluloid (by mixing solid nitrocellulose and camphor) [22, 24]. Nowadays, nitrocellulose is still being used due to its properties—a modulable plastic, lacquer, energetic binder and biomolecular binding membrane. Degree of nitration, the degree of substitution of hydroxyl groups with nitrate esters, is one of the key parameters that affects the performance of a cellulose product [25]. For example, nitrocellulose with a high degree of nitration (>12.5%) is used in explosives [26].

The first semi-synthetic plastic, galalith (the milk stone), was invented in 1897. It was also an accidental discovery, but this time caused by a cat (of Adolf Spittler). The cat spilled formaldehyde into its milk bowl. The milk curdled into a solid material (reaction between casein (of milk) and formaldehyde). It was popular especially for the production of buttons until 1930s [27]. In 1907, Leo Baekeland invented the first fully synthetic plastic—Bakelite (Fig. 1.3). It was marketed as "the material of a thousand uses" due to its unique properties, such as heat resistance [24]. After Hyatt and Baekeland, major chemical companies invested in developing new polymers leading to the expansion of plastic industry. For example, Nylon, a synthetic silk, was invented by Wallace Carothers (DuPont) in 1935 (Fig. 1.3) [24]. Merriam-Webster dictionary defines plastic as any of numerous organic synthetic or processed materials that are mostly thermoplastic or thermosetting polymers of high molecular weight and that can be made into objects, film or filaments [28]. Since 1960s there are increasing concerns about the environmental and

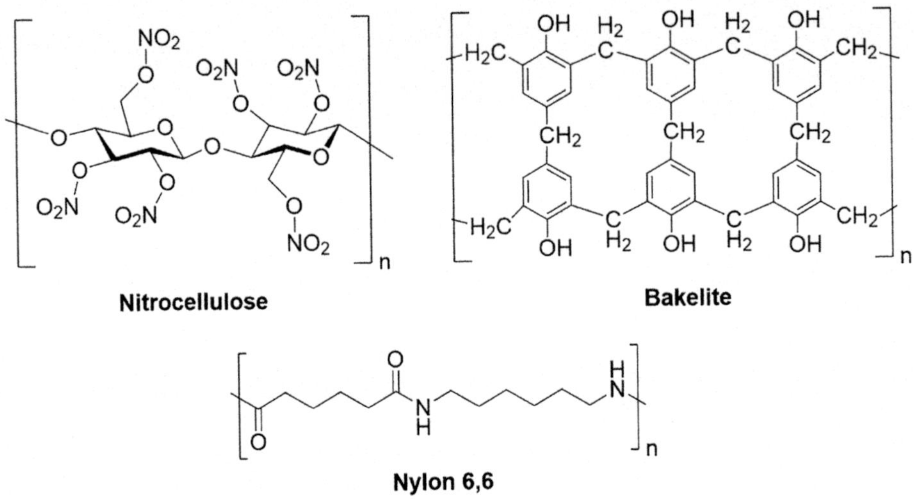

Fig. 1.3 Chemical structure of nitrocellulose, Bakelite, and nylon 6,6

health impacts of plastics, especially those that are synthetic and non-degradable [24]. The word "plastic" gradually became a word used to describe that which was cheap etc., as reminded by the famous quote "I have one word for you: plastics" of the 1967 film—*The Graduate* [24, 29]. This promoted the development of the biopolymer field, particularly that they are biodegradable (e.g., poly(lactic acid) and polyhydroxyalkanoates). Even now, our focus is on nanomaterials and self-assembled materials, and polymers have been widely used to develop these novel materials [30, 31].

When it comes to polymer education and research, Hermann Francis Mark, father of polymer science, developed the world's first academic curriculum in polymer science and technology at the University of Vienna and founded the Polymer Research Institute at the Polytechnic Institute of Brooklyn (the first in the U.S.) [32]. The Journal of Polymer Science was launched in 1946 with Herman as one of the founders [33]. In 1953, Hermann Staudinger was awarded the Nobel Prize in Chemistry for his discoveries in the field of macromolecular chemistry (first related to polymers). He also founded the first polymer journal—Macromolecular Chemistry and Physics (previously *Die Makromolekulare Chemie*) in 1943 [34, 35]. Current journals focusing on polymers and biopolymers are listed in Table 1.1.

Table 1.1 Scientific journals related to polymers and biopolymers

Title	Website
ACS Applied Polymer Materials	https://pubs.acs.org/journal/aapmcd
ACS Macro Letters	https://pubs.acs.org/journal/amlccd
Advances in Polymer Technology	https://onlinelibrary.wiley.com/journal/1631
Biomacromolecules	https://pubs.acs.org/journal/bomaf6
Biopolymers	https://onlinelibrary.wiley.com/journal/10970282
Carbohydrate Polymers	https://www.sciencedirect.com/journal/carbohydrate-polymers
Cellular Polymers	https://journals.sagepub.com/home/crpa
Chinese Journal of Polymer Science	https://www.springer.com/10118
Colloid and Polymer Science	https://www.springer.com/396
Designed Monomers and Polymers	https://www.tandfonline.com/journals/tdmp20
E-Polymers	https://www.degruyter.com/journal/key/epoly/html
European Polymer Journal	https://www.sciencedirect.com/journal/european-polymer-journal
Express Polymer Letters	http://www.expresspolymlett.com/
Fibers and Polymers	https://www.springer.com/12221
High Performance Polymers	https://journals.sagepub.com/home/hip
International Journal of Biological Macromolecules	https://www.sciencedirect.com/journal/international-journal-of-biological-macromolecules
International Journal of Polymer Analysis and Characterization	https://www.tandfonline.com/journals/gpac20
International Journal of Polymeric Materials and Polymeric Biomaterials	https://www.tandfonline.com/journals/gpom20
International Journal of Polymer Science	https://onlinelibrary.wiley.com/journal/9484
Journal of Applied Polymer Science	https://onlinelibrary.wiley.com/journal/10974628
Journal of Bioactive and Compatible Polymers	https://journals.sagepub.com/home/jbc
Journal of Biomaterials Science, Polymer Edition	https://www.tandfonline.com/journals/tbsp20
Journal of Inorganic and Organometallic Polymers and Materials	https://www.springer.com/10904
Journal of Macromolecular Science, Part A: Pure and Applied Chemistry	https://www.tandfonline.com/journals/lmsa20

(continued)

Table 1.1 (continued)

Title	Website
Journal of Macromolecular Science, Part B: Physics	https://www.tandfonline.com/journals/lmsb20
Journal of Polymer Engineering	https://www.degruyter.com/journal/key/polyeng/html
Journal of Polymers and the Environment	https://www.springer.com/10924
Journal of Polymer Research	https://www.springer.com/10965
Journal of Polymer Science	https://onlinelibrary.wiley.com/journal/26424169
Macromolecules	https://pubs.acs.org/journal/mamobx
Macromolecular Bioscience	https://onlinelibrary.wiley.com/journal/16165195
Macromolecular Chemistry and Physics	https://onlinelibrary.wiley.com/journal/15213935
Macromolecular Materials and Engineering	https://onlinelibrary.wiley.com/journal/14392054
Macromolecular Rapid Communications	https://onlinelibrary.wiley.com/journal/15213927
Macromolecular Research	https://www.springer.com/13233
Polymer	https://www.sciencedirect.com/journal/polymer
Polymers	https://www.mdpi.com/journal/polymers
Polymers for Advanced Technologies	https://onlinelibrary.wiley.com/journal/10991581
Polymer Bulletin	https://www.springer.com/289
Polymer Chemistry	http://pubs.rsc.org/en/journals/journalissues/py
Polymer Composites	https://4spepublications.onlinelibrary.wiley.com/journal/15480569
Polymer Degradation and Stability	https://www.sciencedirect.com/journal/polymer-degradation-and-stability
Polymer Engineering and Science	https://4spepublications.onlinelibrary.wiley.com/journal/15482634
Polymer international	https://scijournals.onlinelibrary.wiley.com/journal/10970126
Polymer Journal	https://www.nature.com/pj/
Polymers and Polymer Composites	https://journals.sagepub.com/home/ppc
Polymer Reviews	https://www.tandfonline.com/journals/lmsc20

(continued)

Table 1.1 (continued)

Title	Website
Polymer Testing	https://www.sciencedirect.com/journal/polymer-testing
Progress in Polymer Science	https://www.sciencedirect.com/journal/progress-in-polymer-science
Reactive and Functional Polymers	https://www.sciencedirect.com/journal/reactive-and-functional-polymers

Note This is by no means an exhaustive list and is subject to updates and additions

1.3 Molecular Structure and Types of Polymers and Biopolymers

Molecular structure is the key to understand the science and engineering of polymers [36]. A monomer is a molecule which can undergo polymerization (i.e., the process of converting a monomer or a mixture of monomers into a polymer), thereby contributing constituent units to the essential structure of a macromolecule. Linking monomers does not necessarily lead to polymers. If the molecule is of intermediate relative molecular mass, it is considered as an oligomer molecule. There are different types of polymers (Fig. 1.4), such as homopolymers (derived from one species of monomer) and copolymers (derived from more than one species of monomer). Even for copolymers, there are different types: random copolymers, alternating copolymers (two species of monomeric units distributed in alternating sequence), periodic copolymers (monomeric units appear in an ordered sequence), and block copolymers (a block is defined as a portion of a polymer molecule in which the monomeric units have at least one constitutional or configurational feature absent from the adjacent portions). Polymers can also be linear, branched (a branch or side chain is defined as an oligomeric or polymeric offshoot from a macromolecular chain), and cross-linked (a small region in a macromolecule from which at least four chains emanate, and formed by reactions involving sites or groups on existing macromolecules or by interactions between existing macromolecules) [3, 37]. It is important to point out that the monomeric unit and repeating unit are not the same. Monomeric unit is defined as the largest constituent unit contributed by a single monomer molecule to the structure of a macromolecular or oligomer molecule, while a constituent repeating unit is the smallest constituent unit, the repetition of which constitutes a regular macromolecule, a regular oligomer molecule, a regular block, or a regular chain [3]. There are typically two ways to show a polymer structure: show a few repeating units or include a repeating unit within [] or () combined with a subscript, such as n (Fig. 1.5) [38].

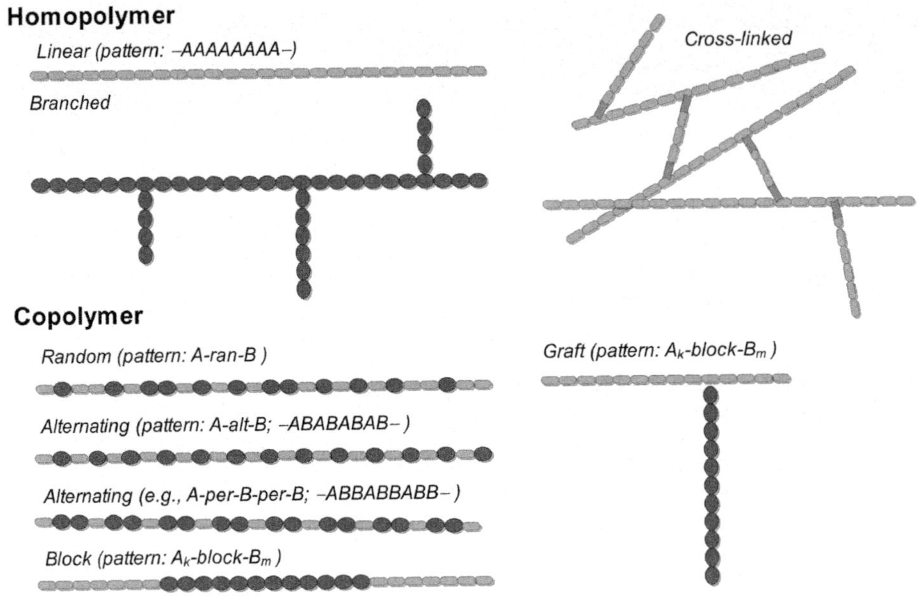

Fig. 1.4 Different types of polymeric structures

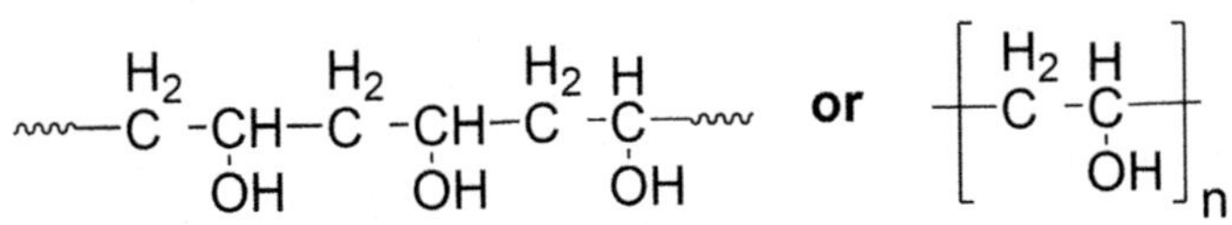

Fig. 1.5 Chemical structure of poly(vinyl alcohol)

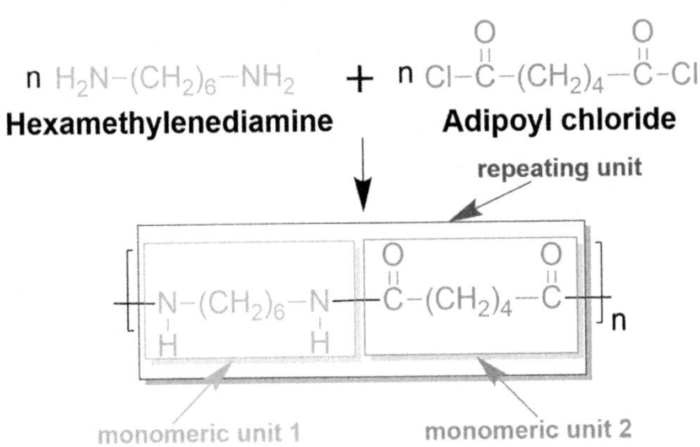

Fig. 1.6 Monomeric units and the repeating unit of nylon 6,6

Problem 1.1
Show the monomeric unit and repeating unit of nylon 6,6 (monomers: hexamethylenediamine and adipoyl chloride).

Answer There are two monomeric units since there are two monomers involved, but only one repeating unit (Fig. 1.6)—monomeric unit 1 (from hexamethylenediamine) + monomeric unit 2 (from adipoyl chloride) = repeating unit.

A corresponding concept to repeating unit is degree of polymerization (DP) which refers to the number of repeating units that constitute a polymer molecule. The subscript n used for describing a polymer structure represents this DP. The relationship between DP and molar mass (M) is given by Eq. 1.1 [38].

$$M = DP \times M_{r.u.} \tag{1.1}$$

where $M_{r.u.}$ is the formular molar mass of the repeating unit. Since chain length of a polymer product varies generally; hence, the average degree of polymerization is used to reflect the average number of repeating units per polymer molecular.

Problem 1.2
A nylon 6,6 product with a molar mass ~ 200,000 g/mol was synthesized through the polymerization of hexamethylenediamine and adipic chloride (1:1 molar ratio) as shown in Fig. 1.6. What is its product's average degree of polymerization (DP)?

Answer Use Eq. 1.1, $M_{r.u.}$ should be calculated first. As shown in Fig. 1.6, the formula of the repeating unit is $C_{12}H_{22}O_2N_2$. Hence, $M_{r.u.}$ is 226 g/mol.

$$DP = M/M_{r.u.} = 200,000/226 \approx 885.$$

When it comes to polymer nomenclature, polymer names have the prefix "poly" followed by enclosing marks around the rest of the name (order: {[()]}). Italicized qualifiers are used for naming non-liner polymers and copolymers, such as poly(styrene-*co*-isoprene) and poly(buta-1,3-diene)-*block*-poly(ethylene oxide) [37]. Polymers can be named based on their source and structure. Linkage-based names generally reference to a class of polymers (examples are proved in Table 1.2) [37, 39].

Lastly, stereoisomerism should be discussed. An isotactic polymer has only one species of configurational base unit in a single sequential arrangement and a syndiotactic polymer shows an alternation of configurational base units that are enantiomeric; while the molecules of atactic polymer have equal numbers of the possible configurational base units in a random sequence distribution [39] (Fig. 1.7).

Table 1.2 Polymer nomenclature

Method	Example(s)
Source-based (monomer name(s))	Homopolymer: polyethylene Copolymer: poly(styrene-*co*-isoprene)
Structure-based (smallest repeating unit)	Poly[oxy(1-bromoethane- 1,2-diyl)]
Linkage-based	Polyester
Trade names	Teflon (polytetrafluoroethylene)
ISO (international organization for standardization) abbreviations	PS (polystyrene) PP (polypropylene)

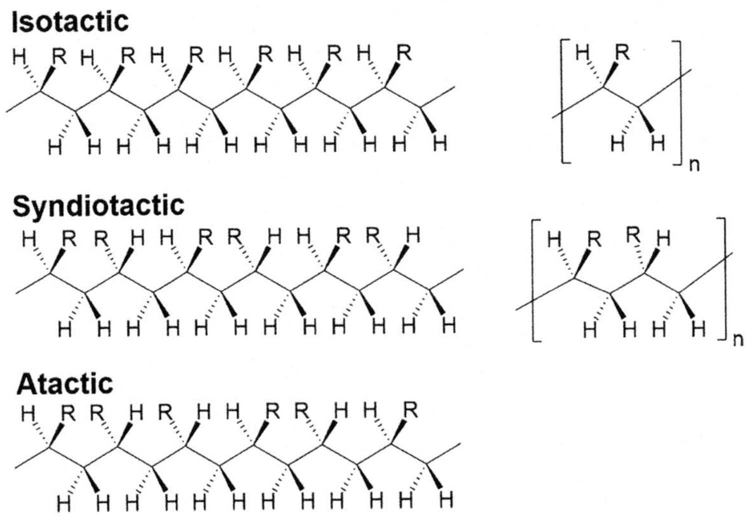

Fig. 1.7 Generalized structures of isotactic, syndiotactic, and atactic polymers

References

1. Jensen WB. The Origin of the Polymer Concept. Journal of Chemical Education. 2008 2008/05/01;85(5):624.
2. Hermann Staudinger and the Foundation of Polymer Science: American Chemical Society. Available from: https://www.acs.org/education/whatischemistry/landmarks/staudingerpolymerscience.html#:~:text=During%20this%20reaction%2C%20which%20he%20called%20%22polymerization%2C%22%20individual,range%20of%20modern%20polymeric%20materials%20and%20innovative%20applications.
3. Jenkins AD, Kratochvíl P, Stepto RFT, et al. Glossary of basic terms in polymer science (IUPAC Recommendations 1996). Pure and Applied Chemistry. 1996;68(12):2287–2311.
4. Hodge P, Hellwich K-H, Hiorns RC, et al. A concise guide to polymer nomenclature for authors

of papers and reports in polymer science and technology (IUPAC Technical Report). Pure and Applied Chemistry. 2020;92(5):797–813.

5. Nagel B, Dellweg H, Gierasch LM. Glossary for chemists of terms used in biotechnology (IUPAC Recommendations 1992). Pure and Applied Chemistry. 1992;64(1):143–168.
6. Smith AM, Moxon S, Morris GA. 13—Biopolymers as wound healing materials. In: Ågren MS, editor. Wound Healing Biomaterials: Woodhead Publishing; 2016. p. 261–287.
7. Biopolymers: Making Materials Nature's Way. Office of Technology Assessment, U.S. Congress; 1993.
8. Endres H-J, Siebert-Raths A. Engineering Biopolymers. Hanser; 2011. (Endres H-J, Siebert-Raths A, editors.).
9. Chen WH, Chen QW, Chen Q, et al. Biomedical polymers: synthesis, properties, and applications. Sci China Chem. 2022;65(6):1010–1075.
10. Kalirajan C, Dukle A, Nathanael AJ, et al. A Critical Review on Polymeric Biomaterials for Biomedical Applications. Polymers (Basel). 2021 Sep 6;13(17).
11. Reddy MSB, Ponnamma D, Choudhary R, et al. A Comparative Review of Natural and Synthetic Biopolymer Composite Scaffolds. Polymers (Basel). 2021 Mar 30;13(7).
12. Hosler D, Burkett SL, Tarkanian MJ. Prehistoric Polymers: Rubber Processing in Ancient Mesoamerica. Science. 1999;284(5422):1988–1991.
13. Goodyear C, Improvement in india-rubber fabrics. patent US3462A. 1844.
14. Somma A. Charles Goodyear and the Vulcanization of Rubber: ConnecticutHistory.org; 2014. Available from: https://connecticuthistory.org/charles-goodyear-and-the-vulcanization-of-rubber/#:~:text=Charles%20Goodyear%E2%80%99s%20discovery%20of%20the%20vulcanization%20of
15. Bin Samsuri A, Abdullahi AA. Degradation of Natural Rubber and Synthetic Elastomers. Reference Module in Materials Science and Materials Engineering: Elsevier; 2017.
16. Cataldo F. Thermochemistry of Sulfur-Based Vulcanization and of Devulcanized and Recycled Natural Rubber Compounds. International Journal of Molecular Sciences. 2023;24(3):2623.
17. Chaikumpollert O, Yamamoto Y, Suchiva K, et al. Mechanical properties and cross-linking structure of cross-linked natural rubber. Polymer Journal. 2012;44(8):772–777.
18. Manaila E, Stelescu MD, Craciun G. Aspects Regarding Radiation Crosslinking of Elastomers. In: Boczkowska A, editor. Advanced Elastomers - Technology, Properties and Applications. IntechOpen; 2012.
19. Yamano M, Yamamoto Y, Saito T, et al. Preparation and characterization of vulcanized natural rubber with high stereoregularity. Polymer. 2021;235:124271.
20. Mostoni S, Milana P, Di Credico B, et al. Zinc-Based Curing Activators: New Trends for Reducing Zinc Content in Rubber Vulcanization Process. Catalysts. 2019;9(8):664.
21. Dochia M, Sirghie C, Kozłowski RM, et al. 2—Cotton fibres. In: Kozłowski RM, editor. Handbook of Natural Fibres. Vol. 1: Woodhead Publishing; 2012. p. 11–23.
22. Nitrocellulose. Encyclopedia Britannica2024. Available from: https://www.britannica.com/science/nitrocellulose
23. Christian Friedrich Schönbein. Encyclopedia Britannica2024. Available from: https://www.britannica.com/biography/Christian-Friedrich-Schonbein
24. Science History Institute. History and Future of Plastics. Available from: https://www.sciencehistory.org/education/classroom-activities/role-playing-games/case-of-plastics/history-and-future-of-plastics/#:~:text=These%20properties%20make%20synthetic%20polymers%20exceptionally
25. Morris E, Pulham CR, Morrison CA. Structure and properties of nitrocellulose: approaching 200 years of research. RSC Advances. 2023;13(46):32321–32333.

26. Fernández de la Ossa MA, Torre M, García-Ruiz C. Nitrocellulose in propellants: characteristics and thermal properties. In: Wythers MC, editor. Advances in Materials Science Research, Volume 7: Nova Science Publishers; 2012. p. 201–220.
27. Guglielmi G. How one cat revolutionized the plastics industry. 2017. Available from: https://www.science.org/content/article/how-one-cat-revolutionized-plastics-industry#:~:text=How%20one%20cat%20revolutionized%20the%20plastics%20industry.%20In%201897,
28. Plastic. Merriam-Webster Dictionary. Available from: https://www.merriam-webster.com/dictionary/plastic
29. Seabrook J. Plastics. The New Yorker. 2010. Available from: https://www.newyorker.com/news/news-desk/plastics
30. Hu S, Yan J, Yang G, et al. Self-Assembled Polymeric Materials: Design, Morphology, and Functional-Oriented Applications. Macromolecular Rapid Communications. 2022;43(14):2100791.
31. Patel GM, Shah V, Bhaliya J, et al. 2—Polymer-based nanomaterials: an introduction. In: Ali N, Bilal M, Khan A, et al., editors. Smart Polymer Nanocomposites: Elsevier; 2023. p. 27–59.
32. Kauffman GB. Herman Francis Mark. Encyclopedia Britannica2024. Available from: https://www.britannica.com/biography/Herman-Francis-Mark
33. Overview. Journal of Polymer Science. Available from: https://onlinelibrary.wiley.com/page/journal/26424169/homepage/productinformation.html
34. Meisel I. As Time Goes By…. Macromolecular Chemistry and Physics. 2003;204(1):13–14.
35. Meisel I, Mülhaupt R. The 60th Anniversary of the First Polymer Journal ("Die Makromolekulare Chemie"): Moving to New Horizons. Macromolecular Chemistry and Physics. 2003;204(2):199–206.
36. Brazel CS, Rosen SL. Fundamental Principles of Polymeric Materials. Wiley; 2012.
37. Hiorns RC, Boucher RJ, Duhlev R, et al. A brief guide to polymer nomenclature (IUPAC Technical Report). Pure and Applied Chemistry. 2012;84(10):2167–2169.
38. Chanda M. Introduction to Polymer Science and Chemistry: A Problem-Solving Approach, Second Edition. Taylor & Francis; 2013.
39. Compendium of Polymer Terminology and Nomenclature: IUPAC RECOMMENDATIONS 2008. The Royal Society of Chemistry; 2009. (Jones RG, Wilks ES, Metanomski WV, et al., editors.).

Polymer Synthesis: Step-Growth and Chain-Growth Polymerization

2

2.1 Introduction to Polymerization

Polymerization is the process of converting a monomer or a mixture of monomers into a polymer. Traditionally, polymerization reactions were classified based on whether or not small molecules are formed, i.e., condensation polymerization and addition polymerization. Polycondensation is the polymerization in which the growth of polymer chains proceeds by condensation reactions between molecules of all degrees of polymerization (with the elimination of small molecules); while polyaddition is the polymerization in which the growth of polymer chains proceeds by addition reactions between molecules of all degrees of polymerization [1]. Polymerization reactions are also be classified based on the reaction mechanisms, i.e., step-growth polymerization and chain-growth polymerization. During step-growth polymerization, the polymer's molar mass increases in a slow, stepwise manner as the reaction time increases. The adjective "chain" of the chain-growth polymerization comes from a chain reaction—in which the growth of a polymer chain proceeds exclusively by the reaction or reactions between a monomer or monomers and a reactive site or reactive sites on the polymer chain with regeneration of the reactive site or reactive sies at the end of each growth step [1, 2]. The differences between step-growth and chain-growth polymerization are summarized in Table 2.1. It is worth noting that there are concerns using these terms as they do not describe the fundamental differences in the polymer growth. For example, both types of polymerization occur in a series of steps and both result in polymer chains [3].

W. Zhang, *Biopolymer Engineering*, Synthesis Lectures on Chemical Engineering and Biochemical Engineering, https://doi.org/10.1007/978-3-031-98868-4_2

Table 2.1 Comparison between step-growth and chain-growth polymerization

Step-growth (e.g., polycondensation)	Chain-growth (e.g., free radical polymerization)
Concentration of monomer(s) decreases rapidly at the beginning	Concentration of monomer decreases steadily throughout the polymerization
Reactions during polymerization: between any two molecules (monomer, oligomer, or polymer)	Reaction during polymerization: monomer + growing chain (with active sites)
Chain growth is exponential	Chain growth is linear (long reaction time is critical for obtaining high molar mass polymer molecules)
Reaction mixture primarily consists of polymer and monomer	Reaction mixture consists of molecules of various chain-lengths

Note Information is from [2–4]

2.2 Step-Growth Polymerization

Step-growth polymerization typically involves one (or more) type(s) of multifunctional monomer(s) where at least bifunctionality is required, with growth occurring between monomers, oligomers, or polymers of any length [3]. There are two general systems: A-B and A-A, B-B. The A-B system involves one type of monomer with two different types of functional groups. As shown in Fig. 2.1., the A-A, B-B system involves two types of monomers. Each type comes with one type of distinct function group (two groups per monomer molecule) [2, 5, 6].

Regarding the A-B system, polylysine is a good example (Fig. 2.1). The Carother's Equation (Eq. 2.1) describe the "degree" of polymerization, X_n (number-average chain length), for a given fractional monomer conversion (or extent of reaction), p $(0 \leq p < 1)$.

$$X_n = \frac{1}{1-p} \tag{2.1}$$

$$p = \frac{N_0 - N}{N_0} \tag{2.2}$$

where N_0 is the initial number of molecules present in the reaction system and N is the number of molecules remaining after certain time [7]. X_n can be used to calculate the number average molar mass $< M_n >$.

$$< M_n >= M_0 \times X_n \tag{2.3}$$

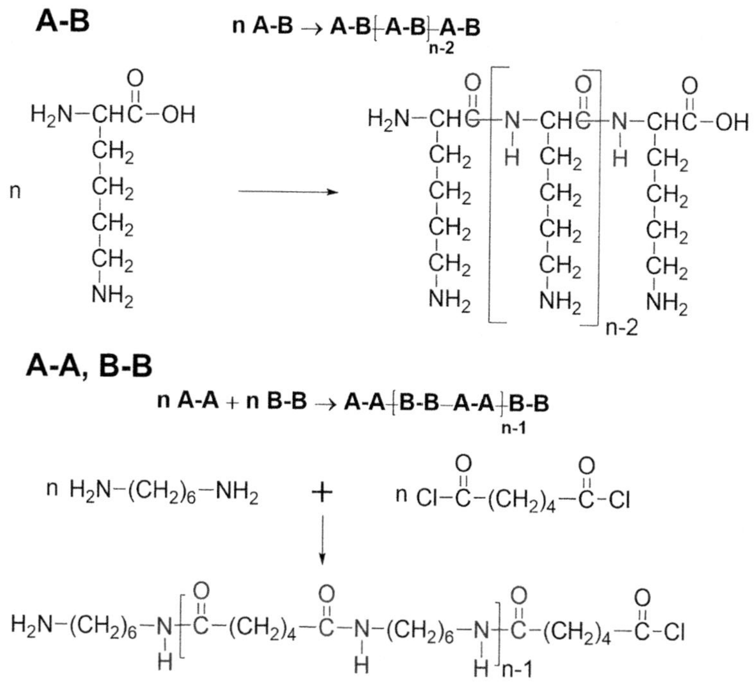

Fig. 2.1 Comparison between A-B and A-A, B-B systems

$$<M_n>= M_0 \times \frac{1}{1-p} \tag{2.4}$$

where M_0 is the molar mass of the monomeric unit. For the A-A, B-B system, half of the molar mass of the repeating unit ($M_{r.u.}$) is used to calculate M_0, since they are two monomeric units [6]. Equation 2.4 is general, which also applies to the A-A, B-B system.

Regarding the A-A, B-B system, nylon 6,6 is a classic example (Fig. 2.1). The Carother's Equation has to be extended (Eq. 2.5).

$$X_n = \frac{1+r}{1+r-2rp} \tag{2.5}$$

where r is the ratio of the number of molecules of the reactants (N_{A0}/N_{B0}; stoichiometric imbalance). The functional group A containing monomer (reactant) is the limiting reactant. Hence, r equals or is less than 1 [5, 6]. As shown in Fig. 2.2, X_n increases with the

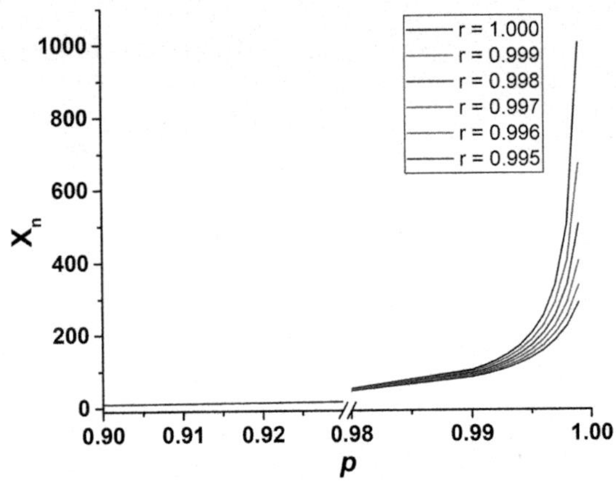

Fig. 2.2 Effects of p and r on X_n

increase of p (at given r). At given p, especially when p is greater than 0.99, the decrease of r leads to significant decrease of X_n (or molar mass of the polymer produced). Hence, to obtain a high molar mass of the polymer product, r should be very close to 1. On the contrary, increasing the input monomer amount could even lead to the decrease of the output polymer's molar mass. Purity of the input monomer is also critical, even a small amount of an impurity can affect the output polymer's molar mass (Problem 2.1) [5, 6]. Considering there are quite a few equations involved, a summary of them is included in Appendix B.

Problem 2.1

During the production of nylon 6,6, equal molar amounts of hexamethylenediamine and adipoyl chloride are intended to be used. However, hexamethylenediamine contains a 3% impurity (by number of molecules). How will this affect the final molar mass of nylon 6,6? $p = 0.99$.

Answer $M_0 = ½(M_{r.u}) = ½ \times 226$ g/mol = 113 g/mol (Problem 1.2).

Without the impurity:

$$r = 1$$

$$X_n = \frac{1 + r}{1 + r - 2rp} = \frac{1 + 1}{1 + 1 - 2 \times 1 \times 0.99} = 100$$

$$< \mathrm{Mn} > = \mathrm{M_0} \times \mathrm{X_n} = 100 \times 113\mathrm{g/mol} = 11,300\mathrm{g/mol}.$$

With the impurity:

$$r = \frac{1 - 0.03}{1} = 0.97$$

$$X_n = \frac{1 + r}{1 + r - 2rp} = \frac{1 + 0.97}{1 + 0.97 - 2 \times 0.97 \times 0.99} \approx 40$$

$$< Mn > = M_0 \times X_n = 40 \times 113\text{g/mol} = 4520\text{g/mol}.$$

With the 3% impurity, the molar mass of nylon 6,6 decreases by ~ 60%.

2.3 Chain-Growth Polymerization

Regarding chain-growth polymerization, free-radical addition polymerization is discussed in this section as an example. Free radicals are independently existing species that possess unpaired electrons and are highly reactive with short lifetimes. Free radical polymerizations are chain polymerizations in which each polymer molecule grows by the addition of a monomer to a terminal free-radical reactive site known as the active center [8]. Free radicals are generated by initiators. An initiator is a molecule that, when subjected to heat, electromagnetic radiation, or chemical reaction, readily undergoes homolytic fission into radicals of greater reactivity than the monomer radical. In general, the decomposition is a first-order kinetic process [6]. Two common types of initiators are peroxide and azo compounds such as azobisisobutyronitrile (Fig. 2.3) [2, 5, 6].

Monomers for chain-growth polymerization generally contain a C = C bond due to the special reactivity of π-bonds to free radicals [6]. A C = C bond can be viewed as a σ + π bond (both Cs are sp^2 hybridized) [9]. There are three stages of a free-radical polymerization process—initiation, propagation, and termination (summarized in Fig. 2.4). During initiation, the active center is created. Propagation is the growth of the polymer chain (one monomer each time). There are different mechanisms involved in

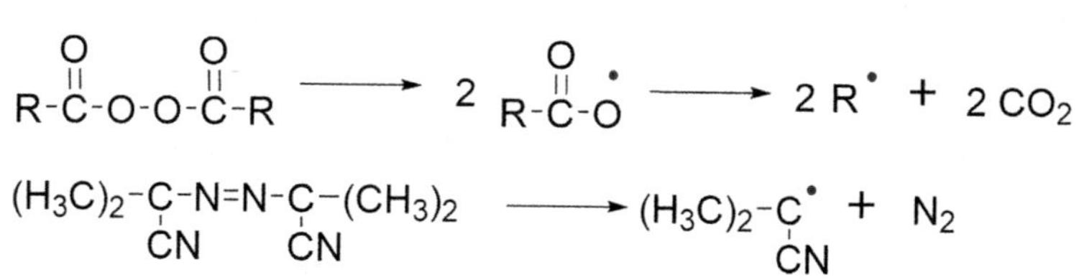

Fig. 2.3 Free radicals' generation from common initiators

Fig. 2.4 Three stages of a free-radical polymerization process

Initiation

I: initiator
M: monomer

$$I \rightarrow 2R^{\bullet}$$

$$R^{\bullet} + M \rightarrow RM^{\bullet}$$

Propagation

$$RM^{\bullet} + M \rightarrow RMM^{\bullet}$$

$$RMM^{\bullet} + M \rightarrow RMMM^{\bullet}$$

$$\downarrow$$

$$RM_{n-1}M^{\bullet} + M \rightarrow RM_nM^{\bullet}$$

Termination

Combination

$$RM_mM^{\bullet} + {}^{\bullet}MM_nR \rightarrow RM_mM\text{-}MM_{n+1}R \rightarrow RM_{m+1}M_{n+1}R$$

$$\text{not } RM_{m+n+2}R$$

Disproportionation

$$RM_mM^{\bullet} + RM_nM^{\bullet} \rightarrow RM_mM' + RM_nM''$$

termination. Two of them are combination and disproportionation. During combination, two chains (mirror images of each other) stick together combining two unpaired electrons to form a single bond. When it comes to the disproportionation mechanism, one chain abstracts a proton from the penultimate carbon on the other chain resulting in two "dead" chains [5, 6].

Problem 2.2

Show the steps of free radical polymerization of vinyl acetate with azobisisobutyronitrile as the initiator.

Answer

Initiation

$I \rightarrow 2R^{\bullet}$

$R^{\bullet} + M \rightarrow RM^{\bullet}$

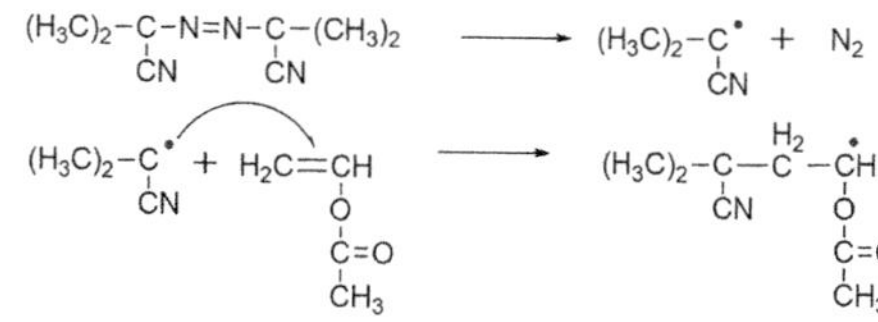

Propagation

$RM^{\bullet} + M \rightarrow RMM^{\bullet}$

$RMM^{\bullet} + M \rightarrow RMMM^{\bullet}$

$RM_{n-1}M^{\bullet} + M \rightarrow RM_nM^{\bullet}$

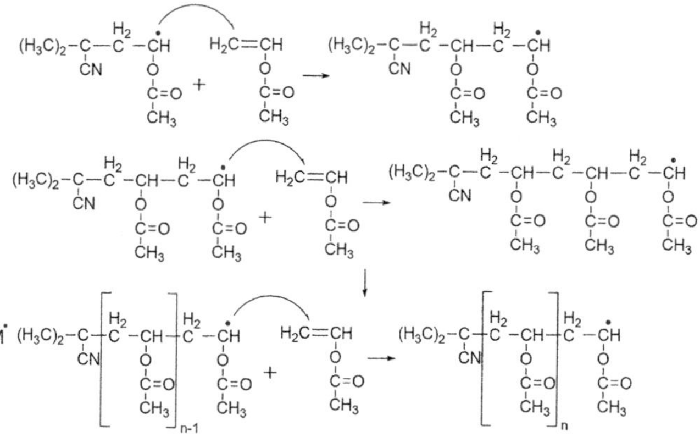

Termination

Combination

$RM_mM^{\bullet} + {}^{\bullet}MM_nR \rightarrow RM_mM\text{-}MM_{n+1}R \rightarrow RM_{m+1}M_{n+1}R$

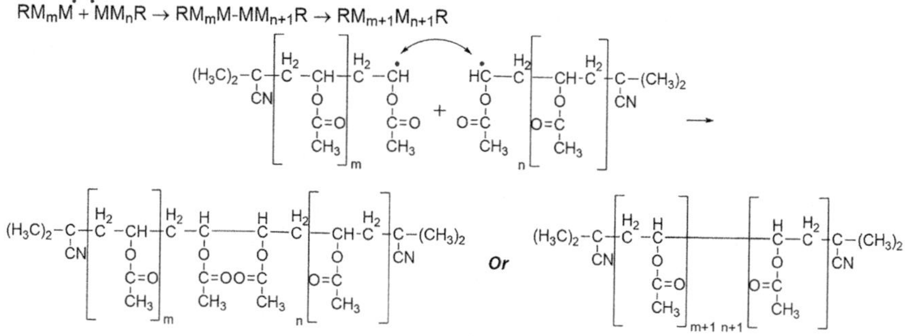

Disproportionation

$RM_mM^{\bullet} + RM_nM^{\bullet} \rightarrow RM_mM' + RM_nM''$

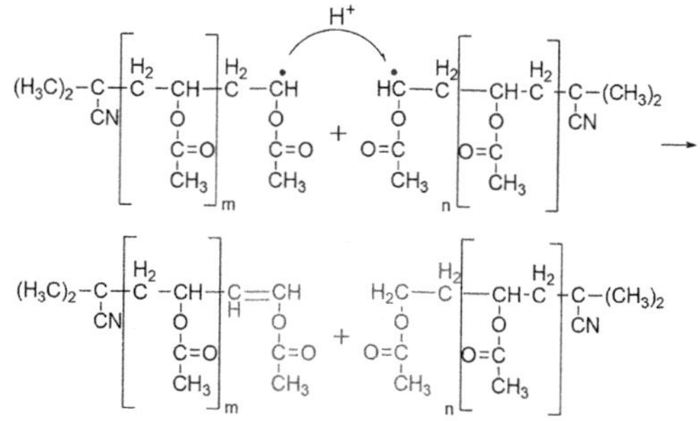

There are four different techniques for free-radical polymerization: bulk, suspension, solution and emulsion. Bulk polymerization only requires monomer and monomer-soluble

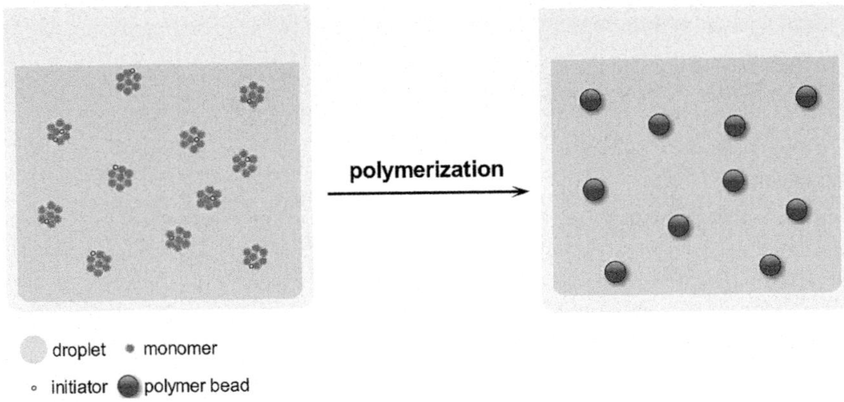

Fig. 2.5 Suspension free-radical polymerization method

initiator (plus other reagents if necessary, such as chain transfer agents for chain length control). Regarding the solution polymerization, monomer and initiator must be soluble in the solvent. The polymer yield (by reaction volume) is lower for solution polymerization than bulk polymerization. On the contrary, water insoluble monomer and initiator are used for suspension polymerization—polymerization occurs within the water droplets containing monomer and initiator (Fig. 2.5). Emulsion polymerization can be used to produce long-chain polymer molecules. The system generally consists of water-insoluble monomer, water, surfactant (emulsification agent), and water-soluble initiator. A surfactant molecule, typically, contains a hydrophilic (polar) head and a hydrophobic (nonpolar) tail. At high enough concentration (reaching the critical micelle concentration; CMC), surfactant molecules form micelles—with polar ends in contact with water and the nonpolar ends inward. As shown in Fig. 2.6, monomers are within micelles formed by surfactant molecules or within droplets "protected" by surfactant molecules. Initiator molecules are water-soluble (limited contact with monomers). Polymerization primarily occurs in micelles due to their large surface area and numbers compared to monomer droplets. Only a very limited number of free radicals can enter the micelles to initiate and terminate the polymerization. An increase in the surfactant concentration alone can increase the polymerization rate and the molar mass of the product (the chain length is a function of the number of micelles in the system at a fixed initiator concentration) [2, 5, 6].

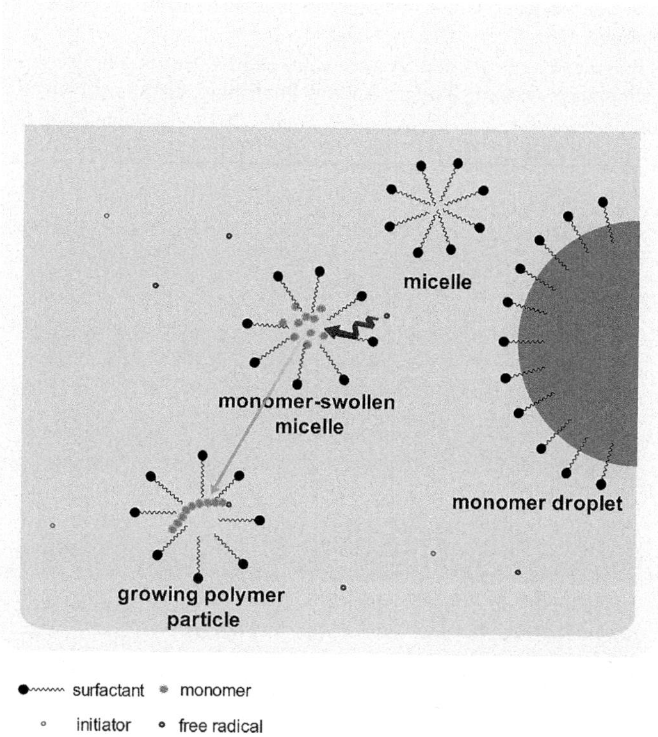

Fig. 2.6 Scheme of an emulsion free-radical polymerization system

References

1. Compendium of Polymer Terminology and Nomenclature: IUPAC RECOMMENDATIONS 2008. The Royal Society of Chemistry; 2009. (Jones RG, Wilks ES, Metanomski WV, et al., editors.).
2. Carraher CE. Introduction to Polymer Chemistry. CRC Press; 2017.
3. Chan CH, Chen J-T, Farrell WS, et al. Reconsidering terms for mechanisms of polymer growth: the "step-growth" and "chain-growth" dilemma. Polymer Chemistry. 2022;13(16):2262–2270.
4. Schaller C. Step Growth and Chain Growth. 2021. Available from: https://chem.libretexts.org/@go/page/238851
5. Brazel CS, Rosen SL. Fundamental Principles of Polymeric Materials. Wiley; 2012.
6. Cowie JMG, Arrighi V. Polymers: Chemistry and Physics of Modern Materials, Third Edition. Informa Healthcare; 2007.

7. Gooch JW. Carothers Equation. In: Gooch JW, editor. Encyclopedic Dictionary of Polymers. New York, NY: Springer New York; 2011. p. 120–121.
8. Obi BE. 2 - Polymer Chemistry and Synthesis. In: Obi BE, editor. Polymeric Foams Structure-Property-Performance. Oxford: William Andrew Publishing; 2018. p. 17–40.
9. Vitz E, Moore JW, Shorb J, et al. Sigma and Pi Bonds. 2023. Available from: https://chem.libretexts.org/@go/page/49306

3 Natural Biopolymers and Biopolymer Manufacturing

3.1 Natural Biopolymers

Three major types of natural biopolymers are polysaccharides, nucleic acids, and proteins (Summarized in Table 3.1). Starch, glycogen, and cellulose are major polysaccharides in biological systems and are D-glucose-based. D (right hand side)- and L (left hand side)-configurations are determined based on the position of the hydroxyl group on the highest numbered chiral carbon from the carbonyl group (C5 in the case of glucose; Fig. 3.1) [1]. In solutions, 99% of glucose exist in two pyranose forms. The epimeric carbon, the carbonyl carbon connected to 2 O atoms, is defined as the anomeric carbon (center). The anomeric carbon is C1 and all the other carbons are numbered clockwise (Fig. 3.1). When it comes to the α and β forms, it is determined by the projection of the hydroxyl group on C1. For the α form, the hydroxyl group projects/points upward which is opposite to the one on C6. For the β form, both hydroxyl groups point in the same direction [2, 3].

Starch is found in plants and is composed of amylose (predominately unbranched) and amylopectin (branched) (Fig. 3.2) [4]. The branched form is insoluble and forms starch granules. Glycogen shows very similar structure to amylopectin but highly branched and is found in animal cells. Different from starch and glycogen, glucoses are linked through β linkages in cellulose. Cellulose is the chief component of plant cell walls. It is unbranched and linear [5]. The aggregated structure of cellulose contains both crystalline (ordered) and amorphous (disordered) regions arranged and stabilized by hydrogen bonds and van der Waals forces between and within cellulose molecules (Fig. 3.3) [6, 7]. Main chains of starch and glycogen are formed by α $(1 \rightarrow 4)$ linkages; while the side chains are linked by α $(1 \rightarrow 6)$ linkages. Cellulose chains are composed of β $(1 \rightarrow 4)$ linkages. Problem 3.1 demonstrates the process for determine the glycosidic linkages of polysaccharides.

W. Zhang, *Biopolymer Engineering*, Synthesis Lectures on Chemical Engineering and Biochemical Engineering, https://doi.org/10.1007/978-3-031-98868-4_3

Table 3.1 Common types of natural biopolymers

Polymer	Monomer	Linkage (covalent bond)	Examples
Polysaccharide	Monosaccharide	Glycosidic	Starch; glycogen; cellulose
Nucleic acid	Nucleotide (nucleoside triphosphate)	Phosphodiester bond	DNA (deoxyribonucleic acid); RNA (ribonucleic acid)
Protein	Amino acid	Peptide bond	Collagen; fibroin

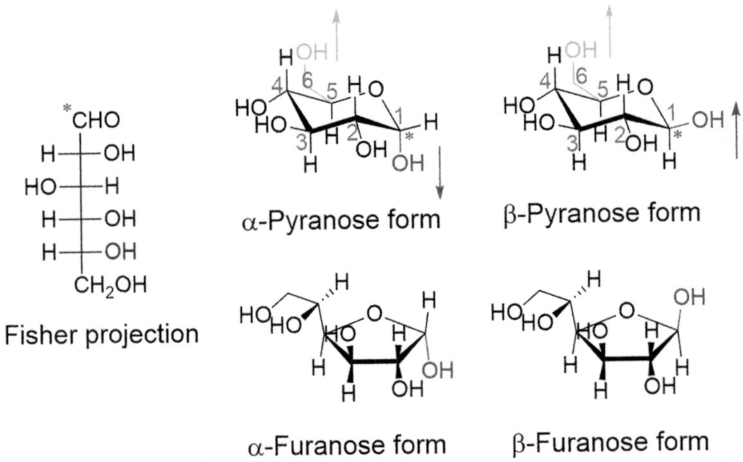

Fig. 3.1 Structure of glucose. * indicates anomeric carbon

Problem 3.1
Determine the glycosidic linkages involved in both the main chain and the branch chain of glycogen.

Answer Main Chain:

a. choose any two connected building units (from left to right)
b. determine the numbers of the carbon involved in the glycosidic linkage formation (C# of the unit on the left → C# of the unit on the left)

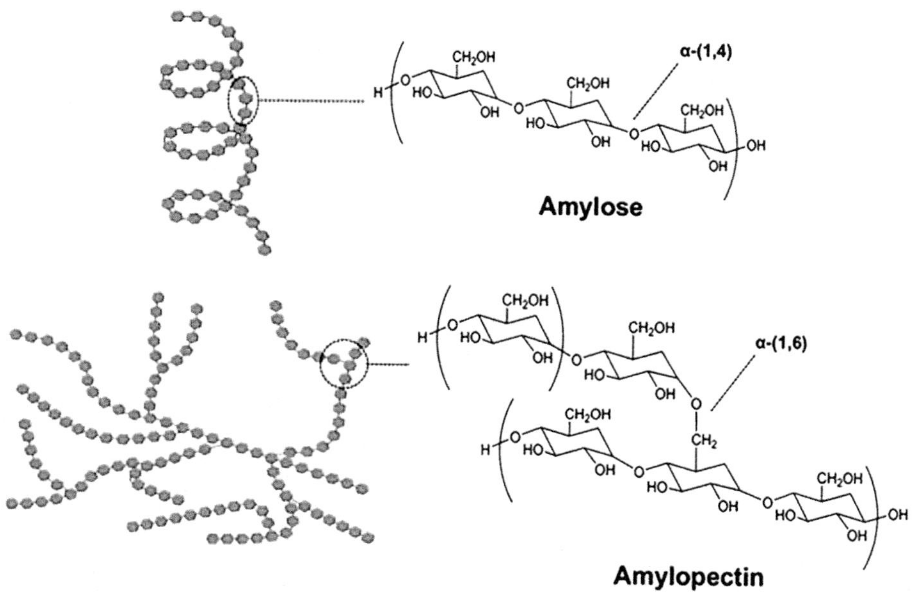

Fig. 3.2 Structure of amylose and amylopectin. *Adapted from* [4] *with permission from the Royal Society of Chemistry*

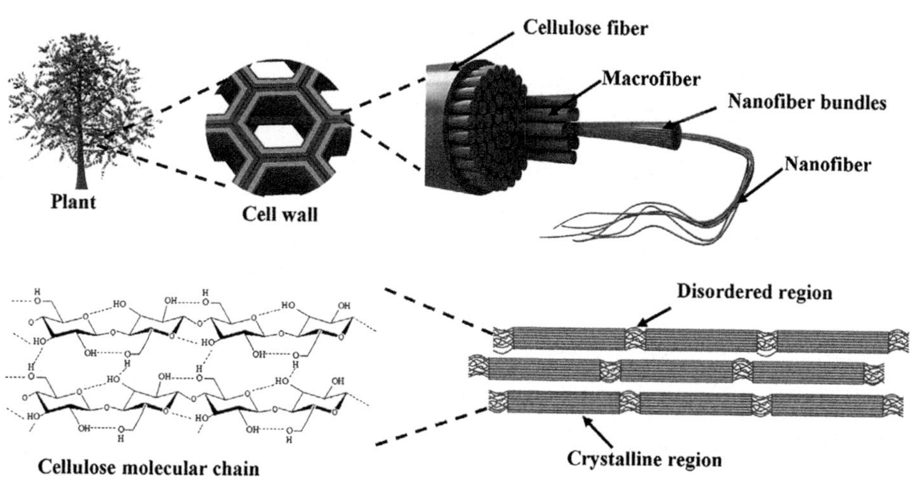

Fig. 3.3 Cellulose hierarchical structure from plant source to molecule level. *Adapted from* [7] *with permission from John Wiley & Sons, Inc.*

c. determine the form (similar to the process demonstrated in Fig. 3.1)—compare the direction of the oxygen of the glycosidic bond to that of C6 (same direction β; opposite direction: α).

Side Chain:

a. choose two connected units (one from the branch and the other from the main chain)
b. determine the numbers of carbon involved (from branch to the main chain)
c. determine the form (same as the main chain).

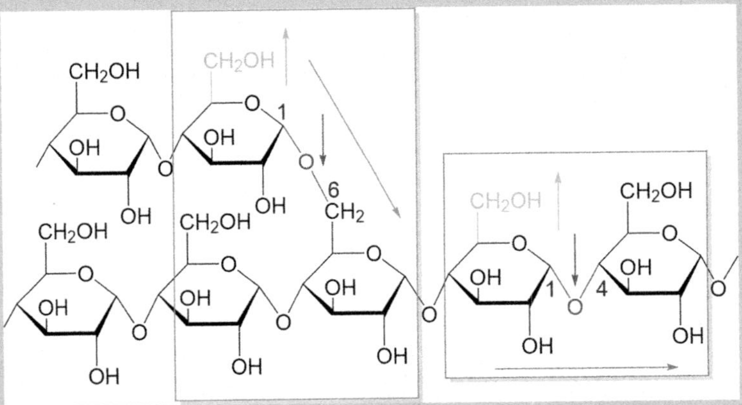

The linkage for the main chain and branched (side chain) is α (1→4) and α (1→6) respectively.

Starch, glycogen, and cellulose are homopolysaccharides. Chitin is another important homopolysaccharide—a polymer of *N*-acetyl glucosamine linked with β (1→4) glycosidic bonds which is very similar to cellulose [8]. In nature, chitin also contains glucosamine units (not abundant) [9]. It is the second most abundant natural polymer and is a structural component of crustaceans, insects, and other arthropods, as well as a component of most fungi and some algae [10]. Chitosan, derived from chitin, is more commonly used due to its unique properties. Chitosan is produced through the deacetylation of chitin converting some of the of *N*-acetyl glucosamine moieties into glucosamine units (Fig. 3.4). It is the only cationic polysaccharide due to the large amount of -NH_2 (-NH_3^+) groups [11].

Heteropolysaccharides are also common and critical in biological systems, e.g., alginate and pectin [12]. Alginate, typically obtained from brown seaweed, is family of linear copolymers of β-D-mannuronate (M) and α-L-gluronate (G) residues. Divalent cations, such as Ca^{2+}, Ba^{2+}, and Sr^{2+}, can link alginate molecules together (i.e. through ionic

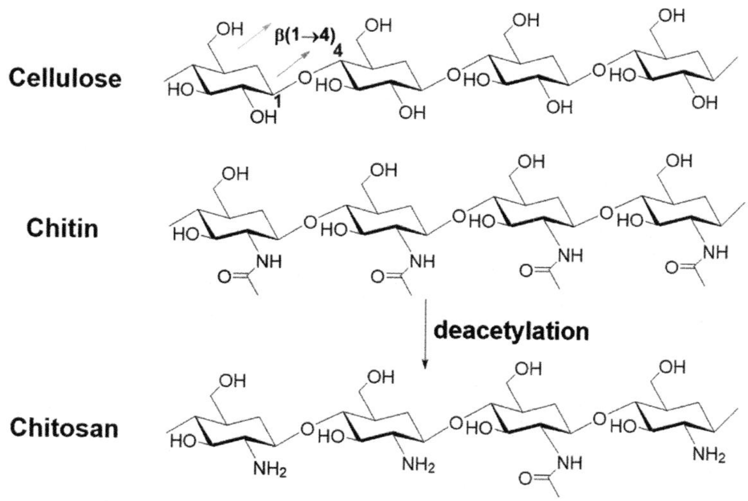

Fig. 3.4 Structure of cellulose, chitin, and chitosan

cross-linking) forming alginate hydrogels [13]. The strong interactions occur between the cations and -COO$^-$ groups of alginate and the resulting structure is referred to as the "egg box" (Fig. 3.5). The G units are believed to be responsible for the cross-linking [14]. Pectins are plant cell wall structural polysaccharides composed mainly of galacturonic acid units with variations in composition, structure and molecular weight [15]. Similar to alginate, divalent cations cause crosslinking throughout the pectin molecules resulting in the formation of shifted "egg-box" structures [16]. Pectin consists of fundamental units of α (1–4)-galacturonic acid, which is often acetylated and/or methylated [17]. Based on the degree of esterification (of carboxyl groups), there are two types of pectins: HM (50–80% of esterification) and LM (< 50% of esterification). The most abundant (~ 65%) pectic polysaccharide is homogalacturonan (HG), a linear homopolymer of α-1,4-linked galacturonic acid. The other pectic polysaccharides include rhamnogalacturonan I and II (RG-I and II) and xylogalacturonan (XGA). RG-I (20–35% of pectin) contains a backbone of the disaccharide repeat [-α-D-GalA-1,2-α-L-Rha-1–4-]$_n$ and RG-II (~ 10% of pectin) is composed of an HA backbone and side chains made of 12 different types of sugars with over 20 different linkages. XGA is an HG substituted at *O*-3 with a β-linked xylose [15, 18, 19] (Fig.3.6).

Polysaccharide biosynthesis is very complex. For extracellular polysaccharides, some of them are first synthesized in the cytosol and then extruded by ATP powered transporters. They can also be assembled outside the cell, synthesized and secreted in a single step, or deposited on the cell surface via vesicular trafficking. In general, polysaccharide biosynthesis requires cytosolic nucleotide-activated monosaccharides which can then be assembled into complex carbohydrate structures by specific glycosyltransferases [20].

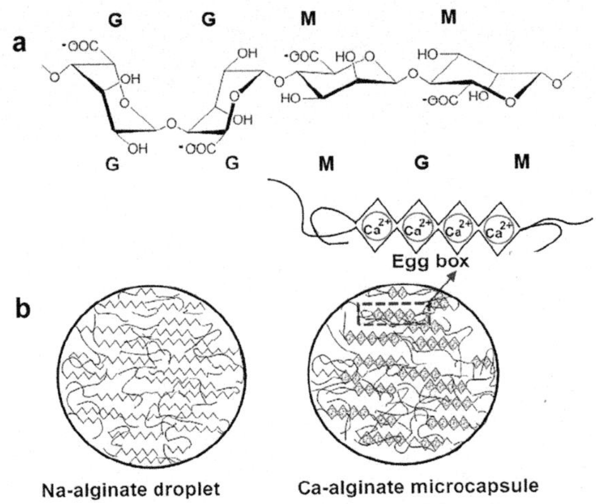

Fig. 3.5 Chemical structure of alginate (**a**) and alginate-based hydrogel formation mechanism (**b**). *Adapted from* [13]

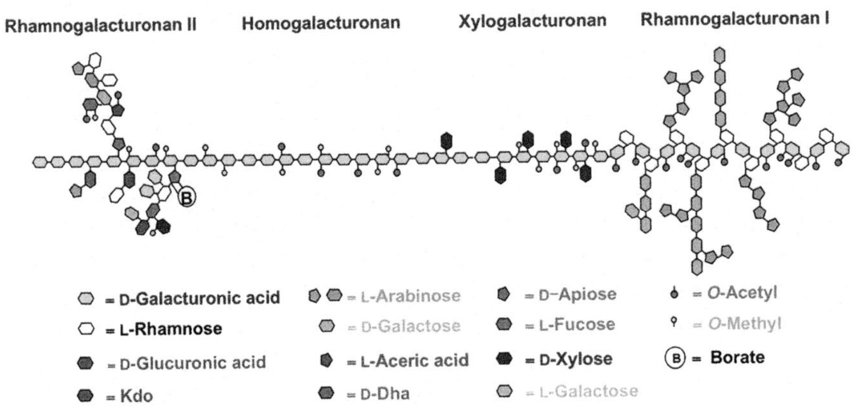

Fig. 3.6 Schematic structure of pectin. *Adapted from* [19] *with permission from the Oxford University Press*

When it comes to chemical synthesis of polysaccharides, there are two types of basic glycans assembly strategies: glycans chain elongation either from the reducing end (with a "free" anomeric carbon not involved in a glycosidic bond) to non-reducing end (Fig. 3.7a) or from the non-reducing end to reducing end (Fig. 3.7b). Hydroxyl groups can be masked by permanent or temporary protecting groups [12, 21].

Nucleic acids, DNA and RNA, are polymers of nucleotides. DNA is the genetic material with the building blocks of deoxynucleoside triphosphates (dATP, dTTP, dCTP, and dGTP). During polymerization, pyrophosphate is released [22]. There are different types

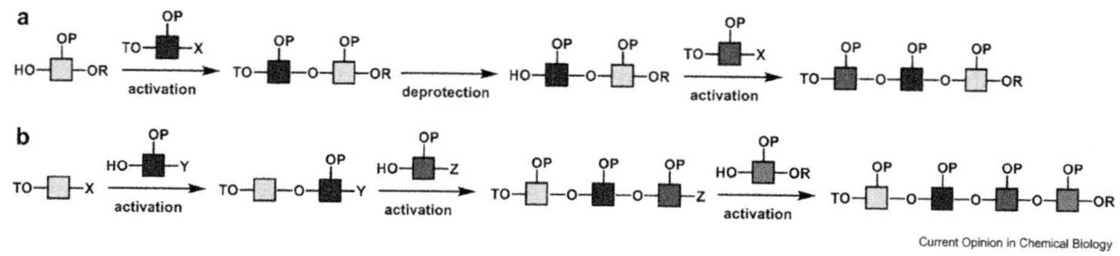

Fig. 3.7 The two basic glycans assembly strategies. P: permanent protecting groups; T: temporary protecting groups; X, Y, Z: leaving groups. *Adapted from* [21] *with permission from Elsevier*

of RNA, such as messenger RNA (mRNA), ribosomal RNA (rRNA) and transfer RNA (tRNA). The nucleotides for RNA contain ribose (deoxyribose for DNA) sugar units and no T bases (but U bases). From the polymer aspect, nucleic acids are polyanions (negatively-charged) due to the phosphate groups [23].

DNA and RNA are synthesized through replication and transcription, respectively, in vivo. In vitro, polymerase chain reaction (PCR) enables amplification (making billions of copies) of a specific DNA fragment from a complex pool of DNA mimicking in vivo replication. As shown in Fig. 3.8 and Table 3.2, PCR requires template DNA, primers, nucleotides (dNTPs; four types of deoxyribonucleoside triphosphates), and DNA polymerase [24]. Synthetic mRNA can be produced by in vitro transcription (IVT) of DNA template using bacteriophage RNA polymerase. IVT reaction requires rNTPs (four types of ribonucleoside triphosphates), RNA polymerase and a DNA template [25]. DNA and RNA can also be synthesized chemically, such as through the phosphoramidite approach. As shown in Fig. 3.9, chain growth proceeds in the 3' to 5' direction (opposite of biosynthesis). The four steps of each cycle are deprotection, condensation (coupling), oxidation, and recapping [26–28].

As biopolymers, nucleic acids, DNA in particular, have been used for different fields, for example as information storage materials. Interestingly, it has been shown that biomass DNA can be directly converted into diverse materials including hydrogels, organogels, composite membranes, plastics through one-step crosslinking (aza-Michael addition; amine group of G attacks the carbon–carbon double bond of poly(ethylene glyco) diacylate (PEGDA) forming nitrogen-carbon bonds) [29]. The same group also reported the

Table 3.2 Comparison between replication and polymerase chain reaction (PCR)

	Replication	PCR
Primer	RNA	DNA
Enzyme(s)	Primase, DNA polymerase, helicase, topoisomerase, ligase	Thermophilic DNA polymerase
Strands separation	Enzyme-controlled	Temperature-controlled
Leading/lagging strand	Yes	No

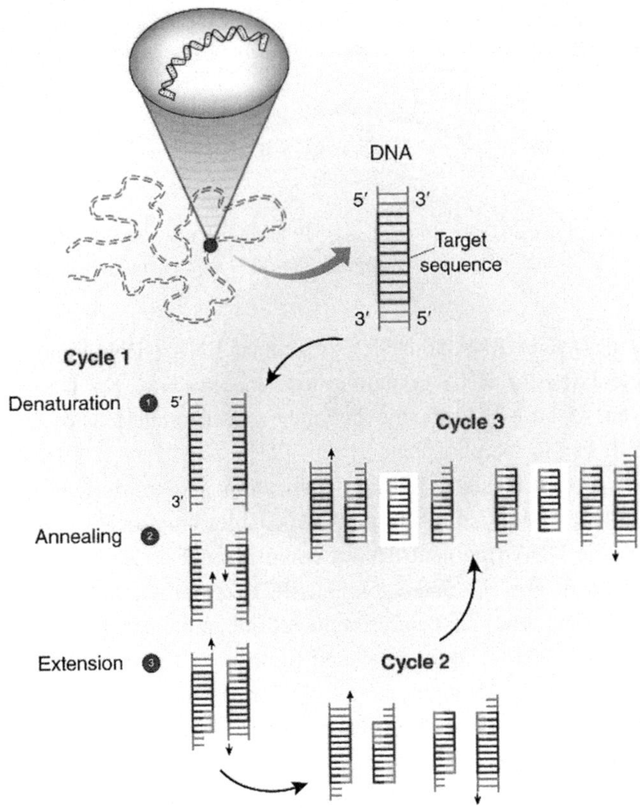

Fig. 3.8 Schematic presentation of the polymerase chain reaction principle. *Adapted from* [24] *with permission from Elsevier*

production of sustainable bioplastics made from biomass DNA and ionomers. Ionomers were used as crosslinkers to form physically cross-linked DNA gels (through electrostatic attraction and hydrophobic interactions). Freeze-drying was then used to process the DNA plastics (Fig. 3.10) [30]. As digital data storage material, DNA exhibits many advantages: ultracompact, stable (if stored properly) and scalable [31–33]. Storing messages in DNA was first demonstrated in 1988 and achieved in 2012 (an html-coded draft of a book including 11 JPG images and 1 JavaScript program) [32, 34]. A typical DNA data storage pipeline consists of five steps: encoding, writing, storing, reading, and decoding [33]. For example, as shown in Fig. 3.11, one bit per base instead of two was encoded (A or C for 0, G or T for 1) in George M. Church et al.'s work, which allowed messages to be encoded many ways in order to avoid sequences that are difficult to read or write, such as extreme GC contents [31]. Since then, different approaches have been developed to

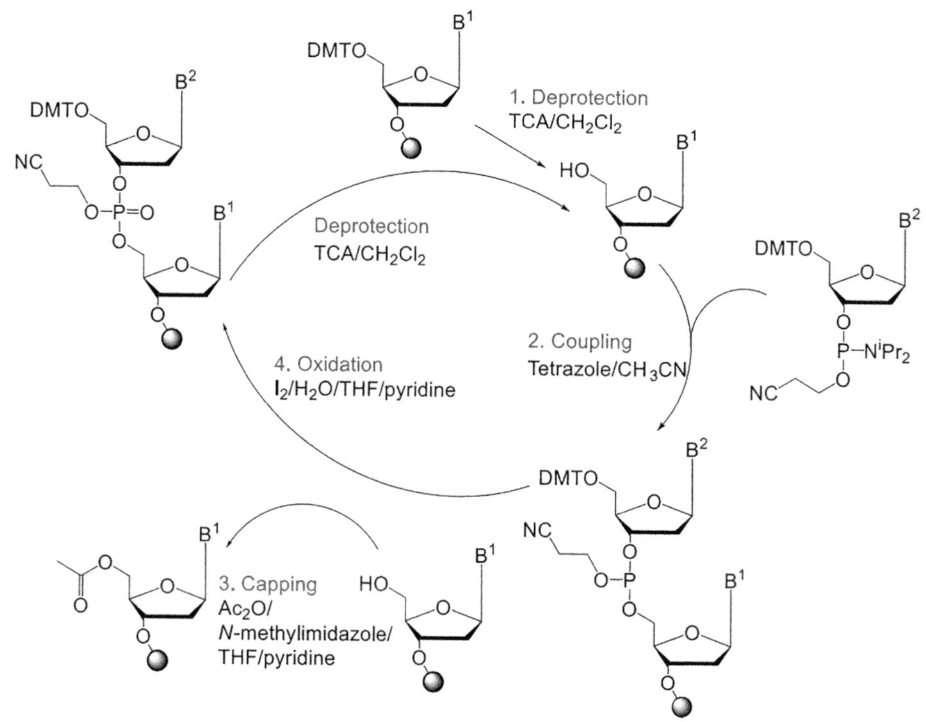

Fig. 3.9 The phosphoramidite approach for oligonucleotide synthesis. DMT: 4,4′-dimethoxytrityl, B: base, TCA: trichloroacetic acid, ^{i}Pr: isopropyl, Ac_2O: acetic anhydride, THF: tetrahydrofuran. *Adapted from* [27] *with permission from Springer Nature*

increase the storage density with the limit considered to be 1.8 bits of data per nucleotide of DNA [32, 33, 35].

DNA-based materials have also been used as biomedical materials due to their biocompatibility, designability & programmability, and excellent selective recognition ability [36–38]. Jong Bum Lee et al., reported a mechanical metamaterial made from a DNA hydrogel. A special bacterial phage polymerase, ϕ 29, was used as it is capable of DNA elongation and displacement. A combination of two unique processes, rolling circle amplification (RCA) and multi-primed chain amplification (MCA), resulting in a DNA hydrogel (Fig. 3.12). The hydrogel showed lipid-like properties when taken out of water (could not keep its shape under the surface tension and gravitational forces) and solid-like properties when in water. The completely deformed hydrogel could return to its original shape after rehydration. Moreover, the controlled drug release using this meta hydrogel was also demonstrated [39]. Michael E. Todhunter et al., reported the DNA programmed synthesis of three-dimensional tissues (Fig. 3.13). This process can generate organoid-like tissues with programmed size, shape, composition and spatial heterogeneity [40]. Other

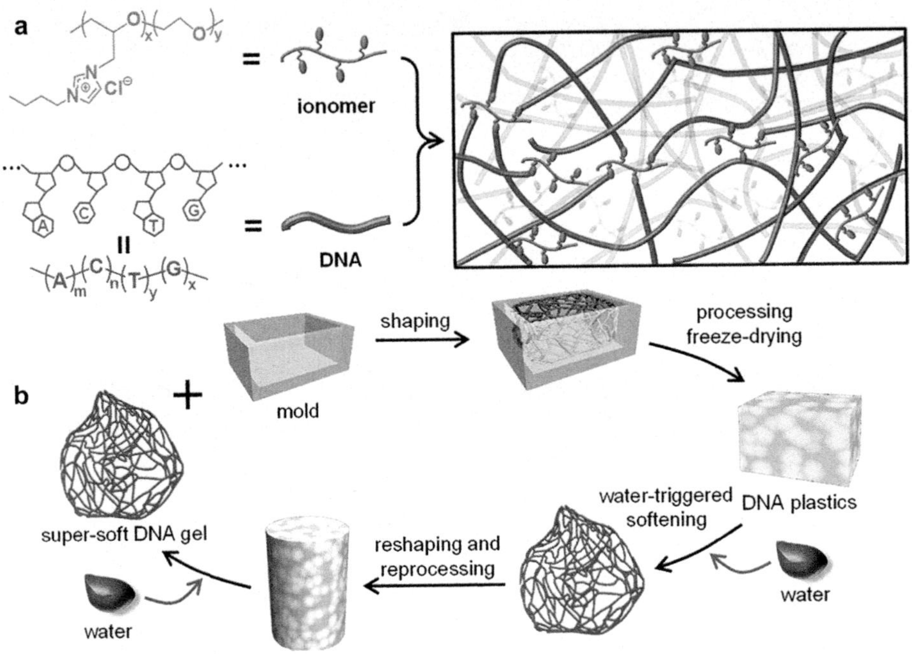

Fig. 3.10 Design and preparation of sustainable DNA plastics. **a** Scheme of the formation of DNA/ionomer hydrogel networks. **b** Scheme of the recyclable use of DNA plastics. *Adapted with permission from* [30]. *Copyright (2021) American Chemical Society*

biomedical applications of DNA-based materials include cell culture [38, 41], targeted drug delivery and multiple drug resistance (MDR) reversion [42, 43], bioprinting [44], biosensing [45], and immunoengineering [46].

All naturally occurring proteins from all living organisms consist of L amino acids [47]. Since both amino acids and proteins contain both acidic and basic groups, they can be net positively-charged, negatively-charged, or neutral. The isoelectric point (pI) of a protein is defined as the pH at which the net charge of a protein molecule is zero. At a pH below their pI, proteins are positively charged. On the contrary, they are negatively charged at a pH above their pI. The effects of pH on –COOH and $–NH_2$ groups are shown in Fig. 3.14. The pI values of proteins have been used for protein isolation, purification and more. For example, isoelectric focusing (IEF) has been used for protein separation. IEF is an electrophoretic technique which separates molecules according to their isoelectric point. In IEF, molecules travel according to their charge under the influence of an electric field, in the presence of a pH gradient, until their net charges become zero (i.e., at the pH which equals to a particle molecule's pI) (Fig. 3.15) [48, 49].

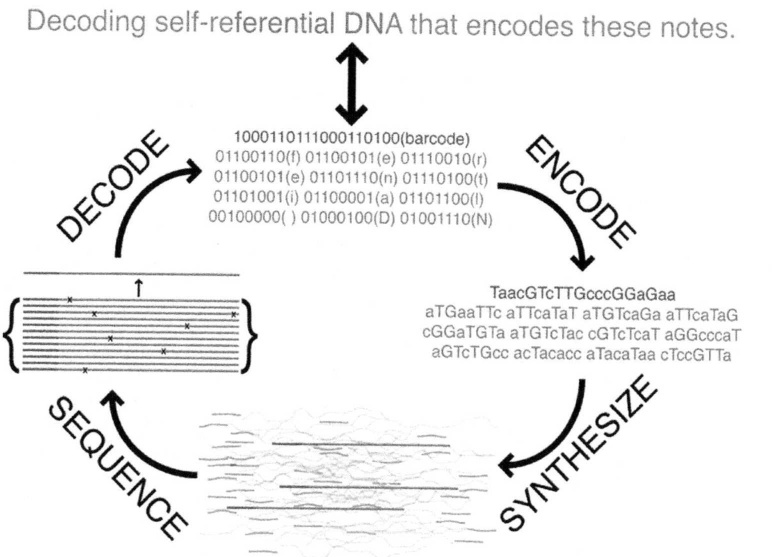

Fig. 3.11 Schematic of DNA information storage. *Adapted from* [31] *with permission from the American Association for the Advancement of Science*

The pI of a protein/peptide can be calculated based the sequence [50]. The equation used is simply.

$$pI = \frac{(pk_{a,1} + pk_{a,2})}{2} \tag{3.1}$$

where $pk_{a,1}$ is the pk_a value of the group which ionization causes the overall molecule charge change from positive to zero; while $pk_{a,2}$ is the pk_a value of the group which ionization causes the overall molecule charge change from zero to positive (increasing the pH gradually from very low pH values). When increasing the pH, –COOH groups become ionized first, generally, and then $–NH_2$ groups. And these groups remain ionized

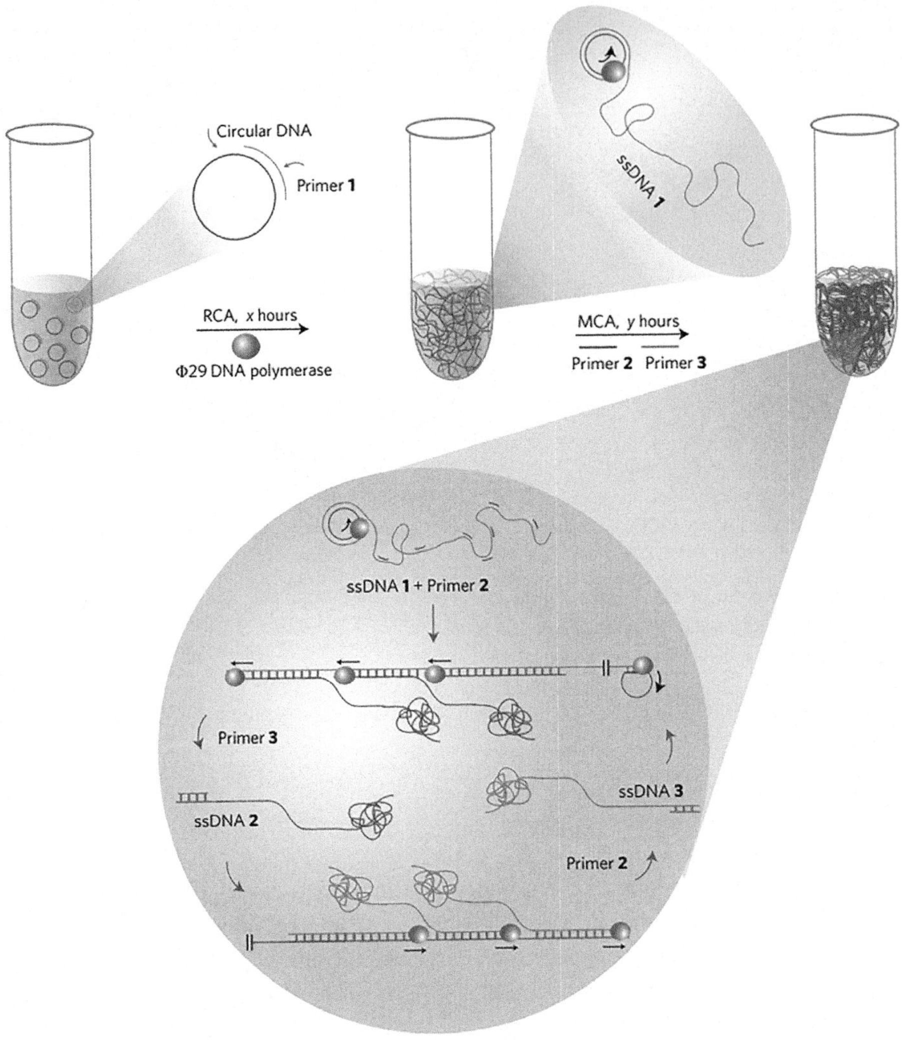

Fig. 3.12 Schematic diagram of the stepwise approach for DNA hydrogel synthesis. ssDNA 1 was first generated using primer 1 through RCA. Then, primers 2 and 3 were used to produce ssDNA 2 (complementary to ssDNA 1) and 3 (complementary to ssDNA 2), respectively through MCA. Thus, ssDNA 1 and 3 had the same sequence and primer 2 could also use ssDNA 3 as the template to synthesize ssDNA 2. RCA: rolling cycle amplification, MCA: multi-primed chain amplification, ssDNA: single-stranded DNA. *Adapted from* [39] *with permission from Springer Nature*

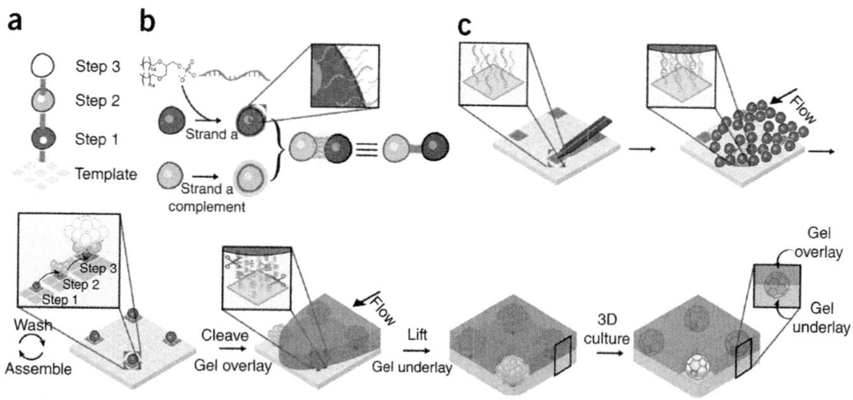

Fig. 3.13 Programming the reconstitution of fully extracellular matrix (ECM)-embedded 3D microtissue arrays by DNA-programmed assembly. **a** Scheme showing the relationship between DNA spots (colored squares), DNA-programmed connectivity (colored lines) and multistep assembly. **b** Incubate cells with lipid-modified oligonucleotides and they can form a temporary chemical adhesion with cells bearing complementary cell-surface oligonucleotides. **c** DNA spots are patterned onto glass slides. Cells bearing complementary cell-surface oligonucleotides are introduced above the patterned substrate and they adhere to the appropriate DNA spot. Then excess cells are removed. Iteration of this process assembles the microtissue into the third dimension. Addition of liquid ECM incorporating DNase releases the assembled microtissues from the template. The gel is peeled off the glass, releasing the tissues. Underlay of the gel with additional ECM results in a fully embedded 3D culture. *Adapted from* [40] *with permission from Springer Nature*

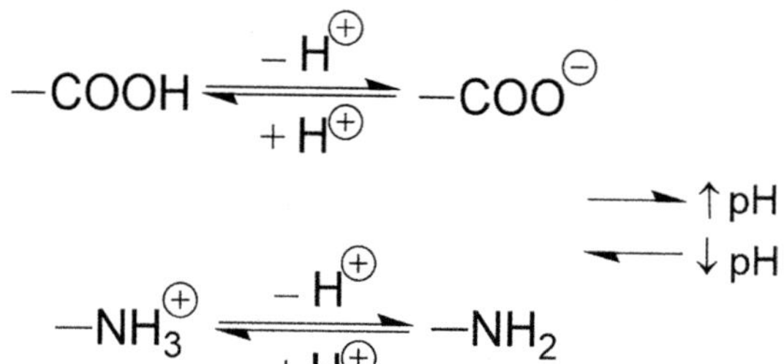

Fig. 3.14 Ionization of carboxyl (through deprotonation) and amino groups (protonation)

if the pH is decreasing. Equation 3.1 is "borrowed" from the equation for calculating the pI of the amino acids without ionizable side (R) groups.

$$pI = \frac{(pk_{a,COOH} + pk_{a,NH_2})}{2} \tag{3.2}$$

where pka, $_{COOH}$ and pka,$_{NH2}$ are the pka values of the –COOH group and –NH_2 linked to the α–carbon (i.e., central carbon), respectively. Equation 3.2 can be derived using the

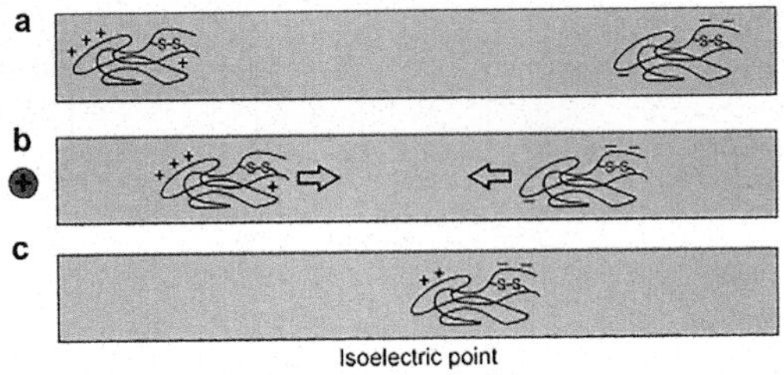

Fig. 3.15 Isoelectric focusing. **a** The same protein placed at various places on the IPG strip, and the charge it obtains (red: low pH; blue: high pH). **b** Protein movement due to the electric field. **c** Protein stops at its isoelectric point. *Adapted from* [49] *with permission from Elsevier*

following Eqs. 3.3–3.6 considering the amino acid as a divalent acid ($H_2A^+ \leftrightarrow HA + H^+ \leftrightarrow A^- + 2H^+$) [51].

$$k_{a,COOH} = \frac{[HA][H^+]}{[H_2A^+]} \tag{3.3}$$

$$k_{a,NH_2} = \frac{[A^-][H^+]}{[HA]} \tag{3.4}$$

$$pk_a = -\log_{10} k_a \tag{3.5}$$

$$pH = -\log_{10}[H^+] \tag{3.6}$$

When pKa = pH, considering a monovalent acid ($HA \leftrightarrow A^- + H^+$), it can be concluded that $[A^-]$ equals [HA] (i.e., half ionized; Eqs. 3.7–3.9). Hence, for calculation purposes, it is easier to consider an ion is fully ionized once the pH is higher than its pk_a.

$$pk_a = -\log_{10}\frac{[A^-][H^+]}{[HA]} = -\log_{10}[H^+] - \log_{10}\frac{[A^-]}{[HA]} = pH - \log_{10}\frac{[A^-]}{[HA]} \tag{3.7}$$

$$pk_a - pH = -\log_{10}\left(\frac{[A^-]}{[HA]}\right) \tag{3.8}$$

$$-\log_{10}\left(\frac{[A^-]}{[HA]}\right) = 0 \rightarrow \frac{[A^-]}{[HA]} = 1 \tag{3.9}$$

Problem 3.2
Estimate the pI of the peptide: Glu-His-Trp-Ser-Gly-Leu-Arg-Pro-Gly.

Answer

a. draw the structure of the peptide (start with $-NH_2$ end)

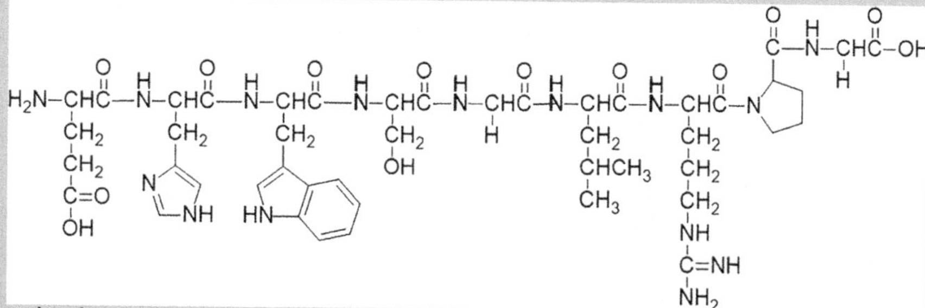

b. write down the pK_a values of all the ionizable groups: left end$-NH_2$ + right end $-COOH$ + all the ionizable R groups

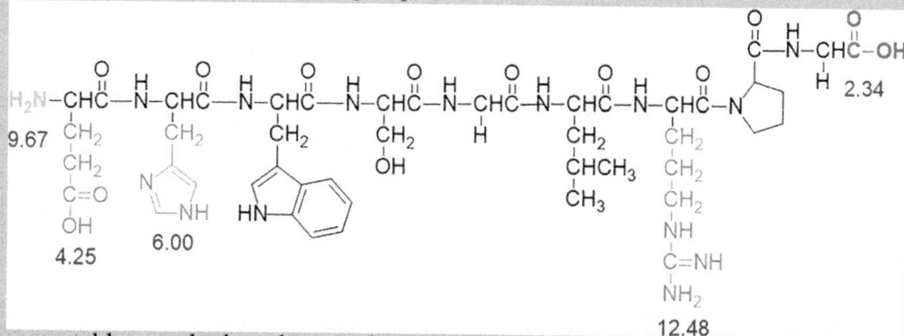

c. use a table to calculate the net charge (gradually increase the pH and top when the net charge becomes negative); $-COOH$ groups only have either negative or no charge and $-NH_2$ groups only exhibit either positive or no charge

pH	pK_a value					Net charge
	Glu-NH_2 9.67	Glu-R 4.25	His-R 6.00	Arg-R 12.48	Gly-COOH 2.34	
1	+	0	+	+	0	+ 3
2	+	0	+	+	0	+ 3
3	+	0	+	+	–	+ 2
4	+	0	+	+	–	+ 2
5	+	–	+	+	–	+ 1
6	+	–	+ /0 (+1/2)	+	–	+ 1/2
7	+	–	0	+	–	0
8	+	–	0	+	–	0
9	+	–	0	+	–	0
10	0	–	0	+	–	– 1

d. determine the two pKa values: pKa,1—the ionization of the corresponding group leads the net charge change from positive to zero and pKa,2—the ionization of the corresponding group leads the net charge change zero to positive

pka1: 6.00

pka2: 9.67

e. calculate the pI

$$pI = \frac{(pk_{a1} + pk_{a2})}{2} = \frac{(6.00 + 9.67)}{2} = 7.84.$$

3.2 Biopolymer Manufacturing

There are different approached for biopolymer manufacturing: chemical synthesis of petro-chemical raw materials, chemical synthesis of bio-based feedstock, direct biosynthesis, and modifications of renewable feedstock [52]. This section focuses on the approaches using natural materials.

Poly(vinyl alcohol) (PVA) is a classic example of a biopolymer produced through chemical synthesis of petro-chemical raw materials. It is water-soluble and biocompatible and has been utilized in various fields [53]. One would assume the monomer for PVA is vinyl alcohol; however, the starting material is vinyl acetate. Vinyl alcohol tautomerizes to acetaldehyde, which is more stable at room temperature. During PVA syntheses, vinyl acetate molecules go through free radical polymerization (emulsion polymerization) first to form poly(vinyl acetate). Then, poly(vinyl acetate) goes through hydrolysis to form PVA (Fig. 3.16) [52–54].

Polylactide (PLA; or poly(lactic acid)) and polyamides are good examples of a biopolymers manufactured through the chemical synthesis of bio-based feedstock [52]. PLA is a biodegradable polyester derived from natural lactic acid from corn, maize, or milk. It can be synthesized either through ring-opening polymerization of the cyclic lactide diester or via direct condensation of lactic acid (Fig. 3.17) [55–57].

Interestingly, CO_2 has been used for biopolymer synthesis. For example, CO_2/epoxide polymerization towards poly(carbonate)s has been widely studied (Fig. 3.18) [58]. Empower Materials, Inc. produces CO_2 polymers such as QPAC®25 poly(ethylene carbonate) and QPAC®40 poly(propylene carbonate) using their proprietary catalyst

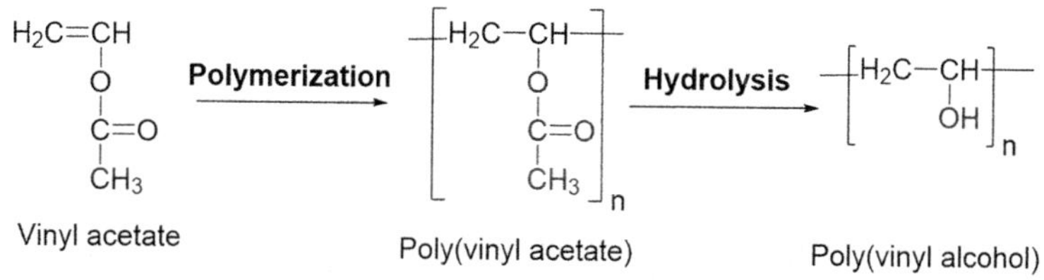

Fig. 3.16 Chemical synthesis process of poly(vinyl alcohol)

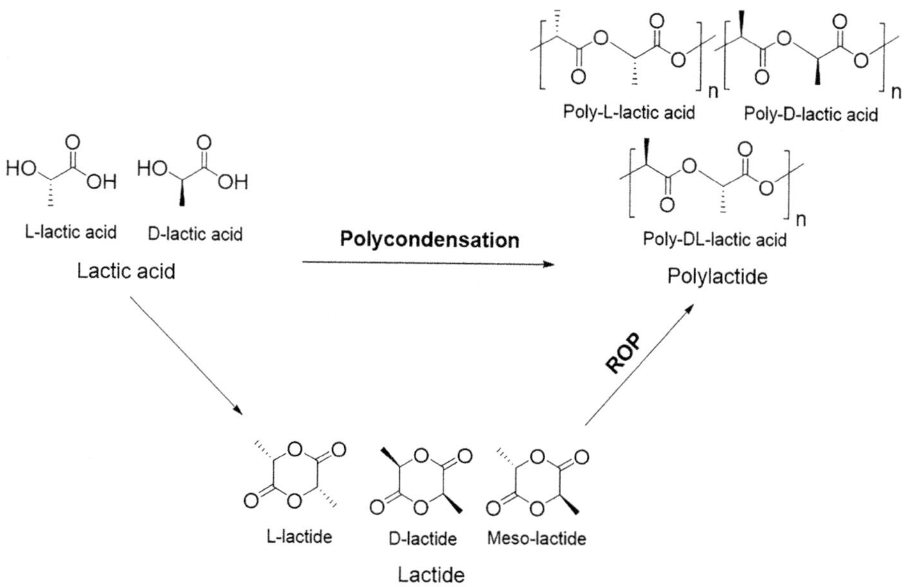

Fig. 3.17 Chemical synthesis process of polylactide. ROP: ring-opening polymerization

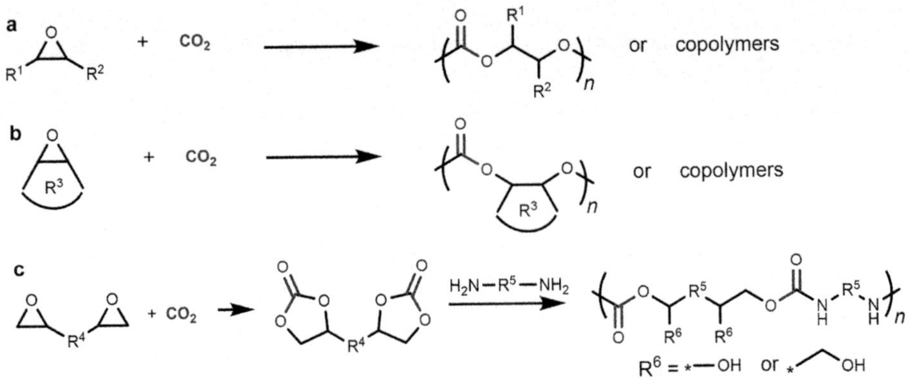

Fig. 3.18 Synthesis of polycarbonates—the synthetic routes toward CO_2 and epoxide-based polymers. *Adapted from* [58] *with permission from Elsevier*

technology [59]. CO has also been used for biopolymer manufacturing commercially. Novomer Inc. produces poly(3-hydroxypropionate). The monomer, beta-propiolactone, is synthesized by reacting ethylene oxide and CO [60].

The field of biopolymer manufacturing through the direct biosynthesis has been growing rapidly, partially due to the advancements in synthetic biology and metabolic engineering. One example is the biosynthesis of natural rubber in dandelion [61]. Commercial production of polyhydroxyalkanoates (PHAs) through direct biosynthesis is of great interest. PHAs are biodegradable and biocompatible and can be produced by many microbes (Fig. 3.19) [62–64]. Newlight Technologies Inc. uses microorganisms found in California to covert air and CO_2 dissolved in saltwater into polyhydroxybutyrate (PHB; belongs to PHAs) [65].

Regarding the biopolymers manufacturing through modifications of renewable feedstock, cellulose and starch derivatives are most common due to their abundance [66]. Take cellulose for example, cellulose has several less favorable properties for application, such as insolubility in water and other common solvents. Various cellulose derivatives

Fig. 3.19 Chemical structure of polyhydroxyalkanoate (PHA) and poly-(*R*)-3-hydroxybutyrate (PHB). *Generated based on* [64]

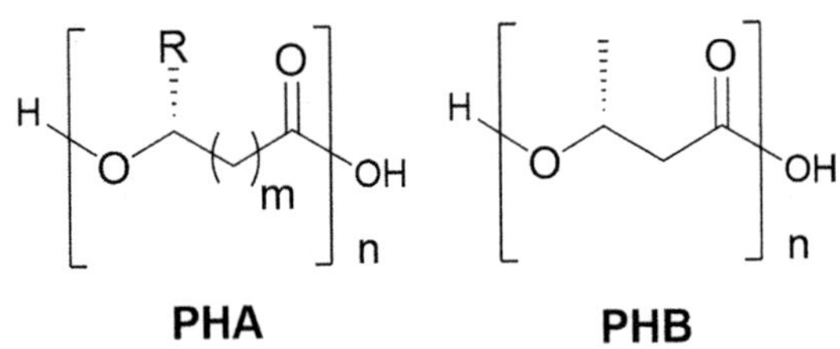

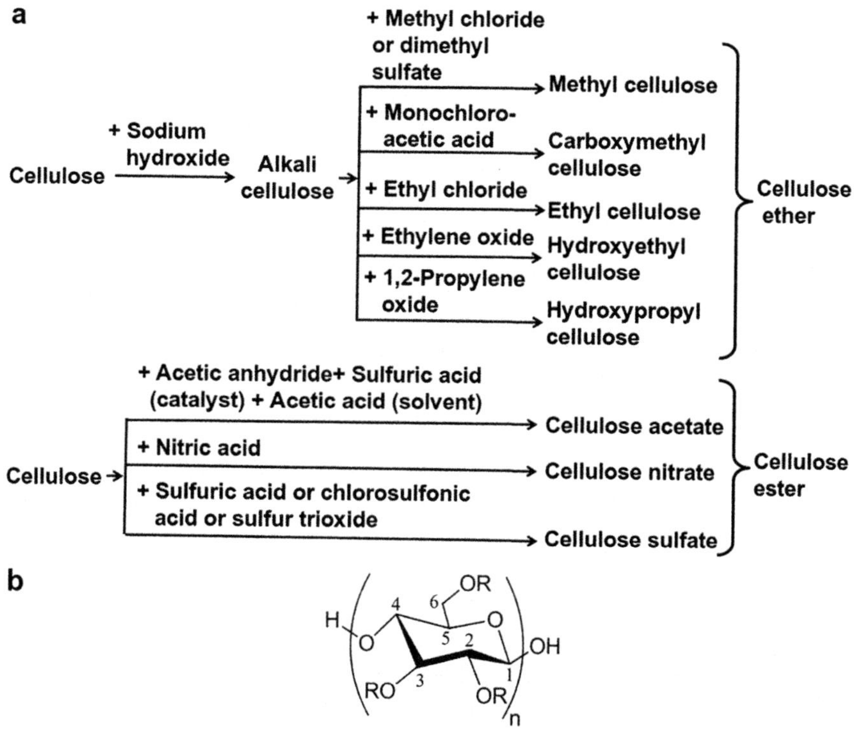

Cellulose derivatives

Cellulose ethers	R	Cellulose esters	R
Methyl cellulose	H or CH_3	Cellulose acetate	H or $(C{=}O)CH_3$
Carboxymethyl cellulose	H or CH_2COOH	Cellulose nitrate	H or NO_2
Ethyl cellulose	H or CH_2CH_3	Cellulose sulfate	H or SO_3H
Hydroxyethyl cellulose	H or CH_2CH_2OH		
Hydroxypropyl cellulose	H or $CH_2CH(OH)CH_3$		

Fig. 3.20 Classification of cellulose derivatives. **a** Chemical conversion procedures of cellulose to its derivatives. **b** Chemical structure of cellulose derivatives. *Adapted from* [67]

have been developed to improve its properties and extend its applications in different fields, such as biomedical industries (Fig. 3.20) [67].

References

1. Shendurse AM, Khedkar CD. Glucose: Properties and Analysis. In: Caballero B, Finglas PM, Toldrá F, editors. Encyclopedia of Food and Health. Oxford: Academic Press; 2016. p. 239–247.
2. Conformations and Cyclic Forms of Sugars. 2015. Available from: https://chem.libretexts.org/@go/page/32568
3. Gunawardena G. Anomeric Carbon. 2018. Available from: https://chem.libretexts.org/@go/page/38592
4. Pereira AGB, Fajardo AR, Valente AJM, et al. Outstanding Features of Starch-based Hydrogel Nanocomposites. In: P M V, Yu L, editors. Starch-based Blends, Composites and Nanocomposites: The Royal Society of Chemistry; 2015.
5. Mason KA, Losos JB, Duncan T. The Chemical Building Blocks of Life. Biology. Thirteenth ed. New York, NY: McGraw Hill; 2023.
6. Poletto M, Pistor V, Zattera AJ. Structural Characteristics and Thermal Properties of Native Cellulose. In: van de Ven TGM, Godbout L, editors. Cellulose - Fundamental Aspects. IntechOpen; 2013.
7. Pei Y, Wang L, Tang K, et al. Biopolymer Nanoscale Assemblies as Building Blocks for New Materials: A Review. Advanced Functional Materials. 2021;31(15):2008552.
8. Fernandes C, Gow NAR, Gonçalves T. The importance of subclasses of chitin synthase enzymes with myosin-like domains for the fitness of fungi. Fungal Biology Reviews. 2016 2016;30(1):1–14.
9. Schmitz C, Auza LG, Koberidze D, et al. Conversion of Chitin to Defined Chitosan Oligomers: Current Status and Future Prospects. Mar Drugs. 2019;17(8).
10. Nunes CS, Philipps-Wiemann P. Chapter 18 - Chitinases. In: Nunes CS, Kumar V, editors. Enzymes in Human and Animal Nutrition: Academic Press; 2018. p. 361–378.
11. Aranaz I, Alcántara AR, Civera MC, et al. Chitosan: An Overview of Its Properties and Applications. Polymers. 2021;13(19):3256.
12. Disaccharides and Glycosidic Bonds. 2021. Available from: https://chem.libretexts.org/@go/page/233984
13. Zhang W. Encapsulation of Transgenic Cells for Gene Therapy. In: Hashad D, editor. Gene Therapy—Principles and Challenges. IntechOpen; 2015.
14. Abka-khajouei R, Tounsi L, Shahabi N, et al. Structures, Properties and Applications of Alginates. Marine Drugs. 2022;20(6):364.
15. Lara-Espinoza C, Carvajal-Millán E, Balandrán-Quintana R, et al. Pectin and Pectin-Based Composite Materials: Beyond Food Texture. Molecules. 2018;23(4):942.
16. Kanungo M, Wang Y, Hutchinson N, et al. Development of Gelatin-Coated Microspheres for Novel Bioink Design. Polymers. 2021;13(19):3339.
17. Tan H, Chen W, Liu Q, et al. Pectin Oligosaccharides Ameliorate Colon Cancer by Regulating Oxidative Stress- and Inflammation-Activated Signaling Pathways. Frontiers in Immunology. 2018;9.
18. Mohnen D. Pectin structure and biosynthesis. Current Opinion in Plant Biology. 2008;11(3):266–277.
19. Harholt J, Suttangkakul A, Vibe Scheller H. Biosynthesis of Pectin. Plant Physiology. 2010;153(2):384–395.
20. Stephens Z, Wilson LFL, Zimmer J. Diverse mechanisms of polysaccharide biosynthesis, assembly and secretion across kingdoms. Current Opinion in Structural Biology. 2023;79:102564.

21. Wang X, Xiao G. Recent chemical synthesis of plant polysaccharides. Current Opinion in Chemical Biology. 2023;77:102387.
22. Minchin S, Lodge J. Understanding biochemistry: structure and function of nucleic acids. Essays Biochem. 2019;63(4):433–456.
23. Roth CM. ELECTROPHORESIS | Nucleic Acids. In: Worsfold P, Townshend A, Poole C, editors. Encyclopedia of Analytical Science (Second Edition). Oxford: Elsevier; 2005. p. 456–460.
24. Garibyan L, Avashia N. Polymerase Chain Reaction. Journal of Investigative Dermatology. 2013;133(3):1–4.
25. Kwon H, Kim M, Seo Y, et al. Emergence of synthetic mRNA: In vitro synthesis of mRNA and its applications in regenerative medicine. Biomaterials. 2018;156:172–193.
26. Roy S, Caruthers M. Synthesis of DNA/RNA and Their Analogs via Phosphoramidite and H-Phosphonate Chemistries. Molecules. 2013;18(11):14268–14284.
27. Brazier J. Chemical Synthesis of Oligonucelotide Sequences: Phosphoramidite Chemistry. In: Scarlett G, editor. DNA Manipulation and Analysis. New York, NY: Springer US; 2023. p. 185–193.
28. Ni S, Yao H, Wang L, et al. Chemical Modifications of Nucleic Acid Aptamers for Therapeutic Purposes. International Journal of Molecular Sciences. 2017;18(8):1683.
29. Wang D, Cui J, Gan M, et al. Transformation of Biomass DNA into Biodegradable Materials from Gels to Plastics for Reducing Petrochemical Consumption. Journal of the American Chemical Society. 2020;142(22):10114–10124.
30. Han J, Guo Y, Wang H, et al. Sustainable Bioplastic Made from Biomass DNA and Ionomers. Journal of the American Chemical Society. 2021;143(46):19486–19497.
31. Church GM, Gao Y, Kosuri S. Next-Generation Digital Information Storage in DNA. Science. 2012;337(6102):1628–1628.
32. Service RF. DNA could store all of the world's data in one room. 2017. Available from: https://www.science.org/content/article/dna-could-store-all-worlds-data-one-room#:~:text=Capable%20of%20storing%20215%20petabytes%20(215
33. Gervasio JHDB, da Costa Oliveira H, da Costa Martins AG, et al. How close are we to storing data in DNA? Trends in Biotechnology. 2024;42(2):156–167.
34. Extance A. How DNA could store all the world's data. Nature. 2016;537(7618):22–24.
35. Erlich Y, Zielinski D. DNA Fountain enables a robust and efficient storage architecture. Science. 2017;355(6328):950–954.
36. Huo S, Kwak M, Qin J, et al. Dynamic DNA-based biomaterials interacting with external, macroscopic, and molecular stimuli. Materials Today. 2021;49:378–390.
37. Jian X, Feng X, Luo Y, et al. Development, Preparation, and Biomedical Applications of DNA-Based Hydrogels. Frontiers in Bioengineering and Biotechnology. 2021;9.
38. Wu R, Li W, Yang P, et al. DNA hydrogels and their derivatives in biomedical engineering applications. Journal of Nanobiotechnology. 2024;22(1):518.
39. Lee JB, Peng S, Yang D, et al. A mechanical metamaterial made from a DNA hydrogel. Nature Nanotechnology. 2012;7(12):816–820.
40. Todhunter ME, Jee NY, Hughes AJ, et al. Programmed synthesis of three-dimensional tissues. Nature Methods. 2015;12(10):975–981.
41. Tang J, Wang J, Ou J, et al. A DNA/Poly-(L-lysine) Hydrogel with Long Shelf-Time for 3D Cell Culture. Small Methods. 2024;8(7):2301236.
42. Morya V, Walia S, Mandal BB, et al. Functional DNA Based Hydrogels: Development, Properties and Biological Applications. ACS Biomaterials Science & Engineering. 2020;6(11):6021–6035.
43. Ma W, Zhan Y, Zhang Y, et al. The biological applications of DNA nanomaterials: current challenges and future directions. Signal Transduction and Targeted Therapy. 2021 2021;6(1):351.

44. Budharaju H, Zennifer A, Sethuraman S, et al. Designer DNA biomolecules as a defined biomaterial for 3D bioprinting application. Materials Horizons. 2022;9(4):1141–1166.
45. Yan C, Hua Y, Guo J, et al. Programmable DNA hydrogels construction with functional regulations for biosensing applications. TrAC Trends in Analytical Chemistry. 2024;173:117628.
46. Maeda M, Kojima T, Song Y, et al. DNA-Based Biomaterials for Immunoengineering. Advanced Healthcare Materials. 2019;8(4):1801243.
47. Structure and Stereochemistry of the Amino Acids. 2020. Available from: https://chem.libretexts.org/@go/page/46018
48. Pergande MR, Cologna SM. Isoelectric Point Separations of Peptides and Proteins. Proteomes. 2017;5(1):4.
49. Drabik A, Bodzoń-Kułakowska A, Silberring J. 7 - Gel Electrophoresis. In: Ciborowski P, Silberring J, editors. Proteomic Profiling and Analytical Chemistry (Second Edition). Boston: Elsevier; 2016. p. 115–143.
50. Tokmakov AA, Kurotani A, Sato KI. Protein pI and Intracellular Localization. Front Mol Biosci. 2021;8:775736.
51. Smidsrød O, Moe S, Moe ST. Biopolymer Chemistry. Tapir Academic Press; 2008.
52. Endres H-J, Siebert-Raths A. Engineering Biopolymers. Hanser; 2011. (Endres H-J, Siebert-Raths A, editors.).
53. Halake K, Birajdar M, Kim BS, et al. Recent application developments of water-soluble synthetic polymers. Journal of Industrial and Engineering Chemistry. 2014;20(6):3913–3918.
54. Hedayati HR, Khorasani M, Ahmadi M, et al. Preparation of well-defined Poly(Vinyl alcohol) by hydrolysis of Poly(Vinyl acetate) synthesized by RAFT suspension polymerization. Polymer. 2022;246:124674.
55. Moreno Raja M, Lim PQ, Wong YS, et al. Chapter 18—Polymeric Nanomaterials: Methods of Preparation and Characterization. In: Mohapatra SS, Ranjan S, Dasgupta N, et al., editors. Nanocarriers for Drug Delivery: Elsevier; 2019. p. 557–653.
56. Ashby MF. Chapter 15—Material profiles. In: Ashby MF, editor. Materials and the Environment (Second Edition). Boston: Butterworth-Heinemann; 2013. p. 459–595.
57. Masutani K, Kimura Y, Chemistry RSo. PLA Synthesis. From the Monomer to the Polymer. In: Jiménez A, Peltzer M, Ruseckaite R, editors. Poly(lactic acid) Science and Technology: Processing, Properties, Additives and Applications: The Royal Society of Chemistry; 2014. p. 0.
58. Song B, Qin A, Tang BZ. Syntheses, properties, and applications of CO_2-based functional polymers. Cell Reports Physical Science. 2022;3(2):100719.
59. Empower Materials. Technology. Available from: https://www.empowermaterials.com/technology
60. Novomer. Product. Available from: https://www.novomer.com/product
61. Epping J, van Deenen N, Niephaus E, et al. A rubber transferase activator is necessary for natural rubber biosynthesis in dandelion. Nature Plants. 2015;1(5):15048.
62. Zhou W, Bergsma S, Colpa DI, et al. Polyhydroxyalkanoates (PHAs) synthesis and degradation by microbes and applications towards a circular economy. Journal of Environmental Management. 2023;341:118033.
63. Thamarai P, Vickram AS, Saravanan A, et al. Recent advancements in biosynthesis, industrial production, and environmental applications of polyhydroxyalkanoates (PHAs): A review. Bioresource Technology Reports. 2022;27:101957.
64. Tullo AH. PHA: A biopolymer whose time has finally come. Chemical & Engineering News. 2019;97(35):20–21.
65. Newlight Technologies. Technology. Available from: https://www.newlight.com/technology

66. Mischnick P, Momcilovic D. Chemical Structure Analysis of Starch and Cellulose Derivatives. In: Horton D, editor. Advances in Carbohydrate Chemistry and Biochemistry. Vol. 64: Academic Press; 2010. p. 117–210.
67. Seddiqi H, Oliaei E, Honarkar H, et al. Cellulose and its derivatives: towards biomedical applications. Cellulose. 2021;28(4):1893–1931.

Polymer and Biopolymer Processing and Characterization

4

4.1 Polymers and Biopolymer Processing

Before they can be used, such as in biomedical applications, polymers and biopolymers have to purified and/or sterilized. Regarding purification, there are four common steps (unit operations) which can be used (not in a particular order):

a. rinse the hydrophilic polymer/biopolymer with organic solvent and remove the organic solvent—remove the hydrophobic impurities dissolved in the organic solvent;
b. dissolve the polymer/biopolymer in a solvent and filter—remove the insoluble impurities;
c. mix polymer/biopolymer solution with activated carbon (charcoal) and filter—activated carbons are carbonaceous materials with highly developed porous structure (large surface area absorption) for removing flavor and odor compounds and more from aqueous solutions [1];
d. precipitate the polymer/biopolymer using its unique properties (such as pH-dependent solubility)—remove the impurities that dissolve in the solvent.

Other steps, such as dialysis, may also be necessary. To design a purification process, the properties of the polymer/biopolymer should be first considered. Other factors should also be considered, such as final purity, cost and time.

W. Zhang, *Biopolymer Engineering*, Synthesis Lectures on Chemical Engineering and Biochemical Engineering, https://doi.org/10.1007/978-3-031-98868-4_4

Problem 4.1

Design a purification process for chitosan (powder).

Answer Start with the properties of chitosan: hydrophilic but only soluble in acidic solutions (depends on its molar mass and such). Since chitosan is a solid, the powder can be rinsed with an organic solvent (such as dichloromethane or chloroform). Then, chitosan can be dissolved in acidic aqueous solution. After treatment with activated carbon, precipitation is followed by increasing the solution's pH (using a strong base of high concentration). Collect the precipitate and wash with deionized water (or dialysis against deionized water). Drying (such as freeze-drying) is the last step [2].

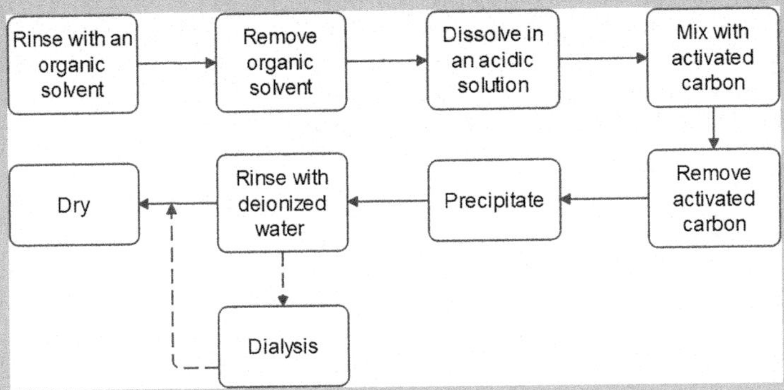

Sterilization is a process of the complete elimination or destruction of all forms of microbial life [3]. When it comes to polymer and biopolymer sterilization, multiple standard approaches are available (summarized in Table 4.1). Like purification, the sterilization method should be chosen based the material properties, time and cost.

Generally, raw polymers and biopolymers are in powder and/or granule form. For actual applications, they have to be processed to form articles of some desired shape and size (i.e., polymer processing) [4]. Some common processing approaches include extrusion, molding, spinning and additive manufacturing. During extrusion, the polymer is fed from a top mounted hopper into the barrel of the extruder. The polymer is melted while being transported by a rotating screw. In the metering zone, the melt becomes homogenized and pressurized and then being pushed through the die (Fig. 4.1) [5, 6].

Molding is commonly used to produce products with a specific shape-design. Injection molding generally consists of two parts: the injection unit and the clamping unit (Fig. 4.2). The polymer/biopolymer granules and such are melted and transported to the screw's tip. Then, the molten polymer/biopolymer is injected into the mold cavity with a constant

Table 4.1 Comparison between common sterilization approaches

Method	Advantage(s)	Disadvantage(s)
High-pressure steam or dry heat	Most effective	Not suitable for heat-/moisture-sensitive polymers/biopolymers
Chemical sterilant (such as ethylene oxide)	Can be used for heat-/moisture-sensitive polymers/biopolymers	Leaves chemical residue; Lead to environmental and health concerns
Physical agents (such as UV light, γ-irradiation)	Can be used for heat-/moisture-sensitive polymer/biopolymers	May cause polymer/biopolymer degradation; Lead to health concerns (accidental operator exposure)
Filtration	Least destructive	Cannot remove all microorganisms; Suitable for small amounts of samples

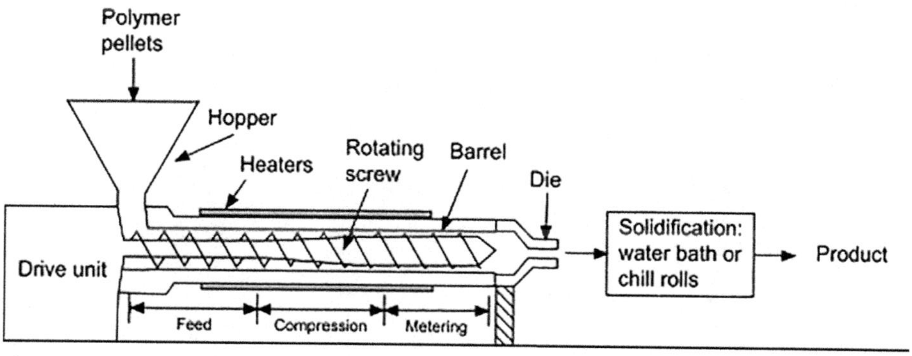

Fig. 4.1 Schematic of a single screw extruder. *Adapted from* [6] *with permission from Elsevier*

volumetric rate. After cooling, the product is ejected [7, 8]. Blow molding is mostly used for producing bottles and such. There are different types of blow molding, such as extrusion blow molding, stretch blow molding, and injection blow molding. Among them, extrusion blow molding is the most common [9]. During blow molding, the mold defines the external shape only, while the inner shape is defined by fluid pressure [10]. During blow molding, the parison (tube of molten polymer/biopolymer) is extruded or dropped into the mold. Then, a pressurized gas is introduced into the parison blowing the heated parison out against the cavity walls (cooled) to form the product (Fig. 4.3) [9, 11].

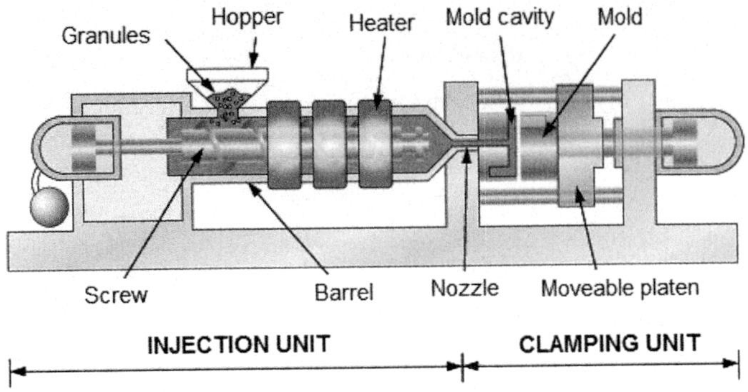

Fig. 4.2 Schematic of an injection molding machine. *Adapted from* [7]

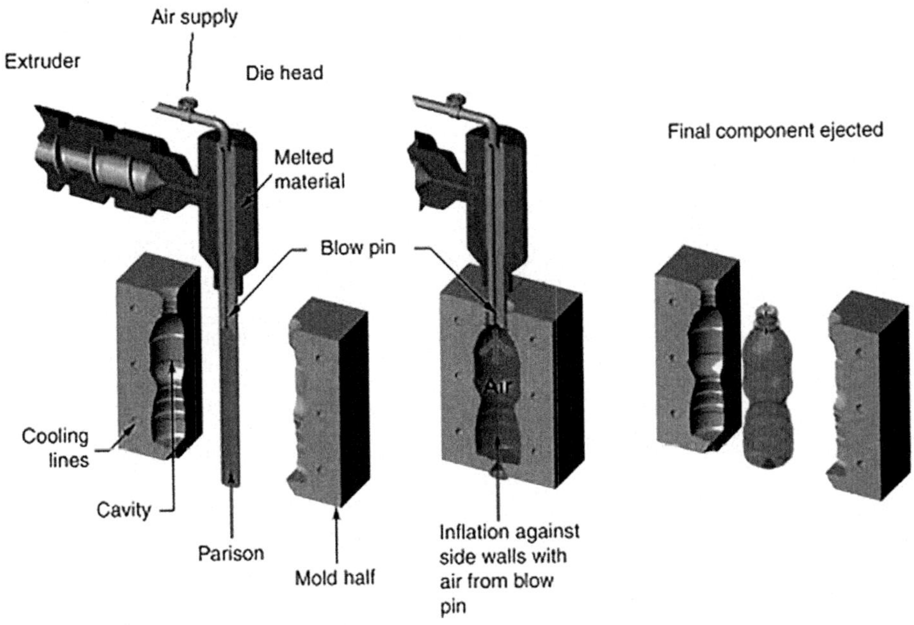

Fig. 4.3 Schematic of extrusion blow molding. *Adapted from* [11] *with permission from Elsevier*

Spinning is used to produce fibers. To produce fibers, the polymer/biopolymer needs to be in a liquid (by dissolving) or semi-liquid (by melting) state. Then the liquid/melt is extruded through a small orifice of a spinneret and followed by hardening (e.g., cooling and coagulation). The three common spinning techniques are wet, dry, and melt spinning (Fig. 4.4) [12, 13]. With the development of nanotechnology, electrospinning has been developed for producing nanoscale fibers. During electrospinning, a viscous polymer/

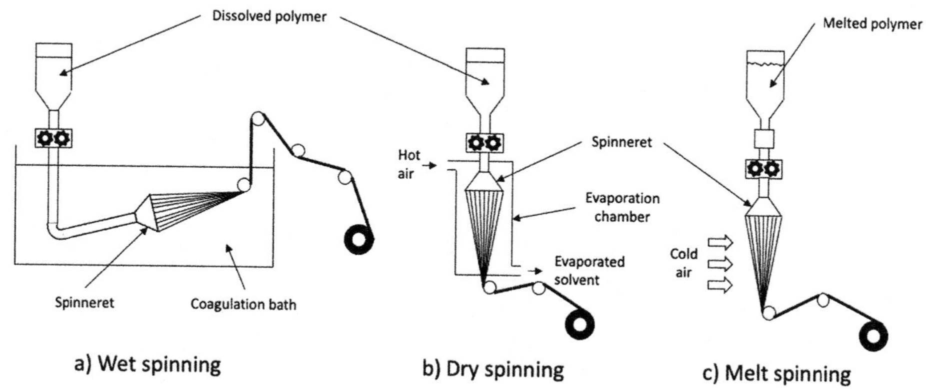

Fig. 4.4 Schematic of different spinning methods. *Adapted from* [13] *with permission from Springer Nature*

biopolymer solution or melt is used. Under high voltage, the pendant droplet deforms into a conical droplet (i.e., the Taylor cone), at the tip of the nozzle (connected to the positive end). At a critical voltage value, as the electrostatic force overcomes the surface tension of the conical droplet, a fine, charged jet is ejected from the tip of the nozzle. The jet undergoes a whipping motion leading to the evaporation of the solvent and stream thinning. The jet solidifies quickly as its diameter decreases, leading to the deposition of solid fiber(s) on the grounded collector [12–14].

Additive manufacturing (or 3D printing) is a process where parts are created layer-by-layer from 3D computer model data (based on digital slicing) [15, 16]. Since introduced in 1980s, additive manufacturing has been widely applied for specialized fabrication, which can create complex geometries and lightweight structures [17, 18]. Other advantages of using additive manufacturing include reduced time-to-market, minimized material waste, and heighted customization [19]. Polymers are the most commonly used material for additive manufacturing. There are a few review articles on polymers/biopolymers involved in additive manufacturing, such as [20, 21]. Materials should be carefully selected (based on their properties) and tested for additive manufacturing. As shown in Fig. 4.5, additive manufacturing techniques can be categorized into four groups: material extrusion, vat photopolymerization, binder jetting, and powder bed fusion [16] (Fig. 4.6).

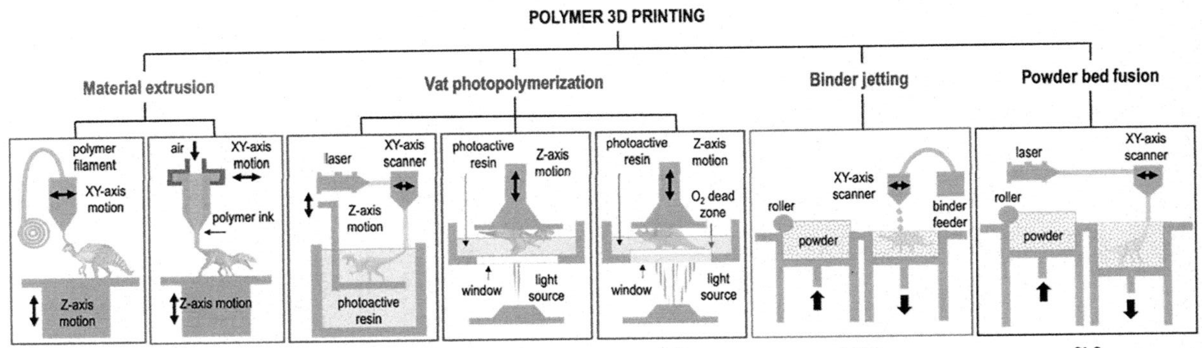

	FDM	DIW	SLA	DLP	CLIP	Inkjet	SLS
Advantages	Cheapest technology, strong parts, variety of nontoxic materials	Relatively cheap, variety of materials, mild print conditions	High resolution, complex geometries, smooth finish	High resolution, complex geometries, smooth finish, fast build speeds	Fast build speeds, high resolution, complex geometries, smooth finish, no "peel" step	Inexpensive, mild print conditions, no support structures	Fast build speeds, relatively low cost, no support structures
Disadvantages	Low resolution, poor surface finish, relatively slow build speeds, need for support structures, poor reproducibility	Relatively slow build speeds, poor mechanical strength, low resolution, limited geometric complexity, need for support structures, poor reproducibility	Post-curing steps, need for support structures, limited range of materials, products prone to shrinkage, slow build speed	Post-curing steps, need for support structures, limited range of materials, products prone to shrinkage, low upper limit for viscosity of ink	Post-curing steps, parts often require support structures, limited range of materials, products prone to shrinkage	Limited reproducibility, poor surface finish, post-processing steps, low upper limit for viscosity of ink, long drying times, poor mechanical strength	Resolution dependent on particle size, powdery surface finish, limited range of materials
Common polymers	PLA, PCL, Polycarbonates, PLGA	PLA, naturally occurring polymers	PPF, Polyacrylates	PPF, Polyacrylates	PPF, Polyacrylates	PCL, PLA/PEG	Polycarbonates, nylons, poly(styrene)

Fig. 4.5 Overview of the common 3D printing (additive manufacturing) techniques used for polymer processing and their advantages/disadvantages. FDM: fused deposition modeling, DIW: direct ink write, SLA: stereolithography, DLP: digital light processing, CLIP: continuous liquid interface production, SLS: selective laser sintering, PLA: polylactide, PCL; polycaprolactone, PLGA: poly(lactic-*co*-glycolic acid)p, PPF: poly(propylene fumarate), PEG: poly(ethylene glycol). *Adapted with permission from* [16]. *Copyright (2021) American Chemical Society*

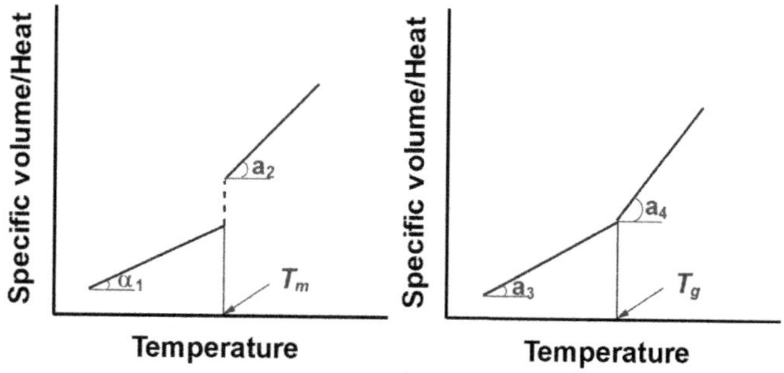

Fig. 4.6 Specific volume/Heat-Temperature plot for crystalline and amorphous polymers. *Generated based on* [32–34]

Additives are commonly used during polymer/biopolymer processing to improve their properties such as stiffness, toughness, and durability. Common additives are plasticizers, stabilizers, impact modifiers, and flame retardants [22]. For example, plasticizers are added to a polymer solution to promote its plasticity and flexibility. Plasticizers lower the glass transition temperature of the polymers. It is hypothesized the plasticizers embed themselves between the individual polymer chains and disintegrate the polymer–polymer interactions (increasing the distance between polymer chains) [22, 23].

4.2 Polymers and Biopolymer Characterization

Regarding the polymer/biopolymer characterization, molar mass is one of the most important parameters. The molar mass of a polymer/biopolymer generally has a significant effect on its rheological properties, electrical conductivity, dielectric strength, surface tension and more [24]. The molar mass of a polymer/biopolymer product is generally an average value as it contains chains of varying lengths. Two commonly used average molar mass are number-average ($<M_n>$) and weight-average ($<M_w>$) molar mass. $< M_n>$ is based on the chain length average, while $<M_w>$ is determined based on chain weight average. For $< M_w >$, longer chains (higher weight (mass)) play a more significant "role" in determining the average molar mass. Hence, for a specific sample, $<M_w>$ is generally higher than $<M_n>$. The ratio of $< M_w>$ to $<M_n>$ is called the polydispersity index (PDI) [25]. The higher the PDI, the broader the chain length variation. The PDI is 1 for monodisperse polymers [26]. $<M_n>$ and $<M_w>$ can be calculated using the Eqs. 4.1 and 4.2, respectively.

$$<M_n> = \frac{\sum N_i M_i}{\sum N_i} = \frac{\sum n_i M_i}{\sum n_i} = \frac{\sum w_i}{\sum \frac{w_i}{M_i}} \tag{4.1}$$

$$<M_n> = \frac{\sum N_i M_i^2}{\sum N_i M_i} = \frac{\sum n_i M_i^2}{\sum n_i M_i} = \frac{\sum w_i M_i}{\sum w_i} \tag{4.2}$$

N_i, n_i, M_i, and w_i is the number, molar amount, molar mass, and mass of molecules of species i, respectively. The relationship of these terms can be expressed as:

$$w_i = \frac{N_i}{N_A} M_i = n_i M_i \tag{4.3}$$

where N_A is the Avogadro's constant (6.022×10^{23} mol^{-1}) [25, 27].

Problem 4.2

A biopolymer sample consists of chains of five distinct molar masses, 1.0×10^5, 2.5×10^5, 4.0×10^5, 7.5×10^5, and 1.0×10^6 g/mol with the molar amount ratio of 1:2:5:3:1. Determine both $<M_n>$ and $<M_w>$.

Answer M_i and n_i are given. Assuming the molar amount of the species with molar mass of 100,000 g/mol is 1 mol and using Eqs. 4.1 and 4.2,

$$M_n = \frac{\sum n_i M_i}{\sum n_i}$$

$$= \frac{\sum(1 \times 1.0 \times 10^5 + 2 \times 2.5 \times 10^5 + 5 \times 4 \times 10^5 + 3 \times 7.5 \times 10^5 + 1 \times 1.0 \times 10^6)\,\text{mol} \times \frac{\text{g}}{\text{mol}}}{\sum(1 + 2 + 5 + 3 + 1)\,\text{mol}}$$

$$= \frac{5.85 \times 10^6 \text{g}}{12\,\text{mol}} = 4.875 \times 10^5 \frac{\text{g}}{\text{mol}}$$

$$<M_n> = \frac{\sum n_i M_i^2}{\sum n_i M_i}$$

$$= \frac{\sum\left[1 \times (1.0 \times 10^5)^2 + 2 \times (2.5 \times 10^5)^2 + 5 \times (4 \times 10^5)^2 + 3 \times (7.5 \times 10^5)^2 + 1 \times (1.0 \times 10^6)^2\right]\text{mol} \times \left(\frac{\text{g}}{\text{mol}}\right)^2}{\sum(1 \times 1.0 \times 10^5 + 2 \times 2.5 \times 10^5 + 5 \times 4 \times 10^5 + 3 \times 7.5 \times 10^5 + 1 \times 1.0 \times 10^6)\,\text{mol} \times \frac{\text{g}}{\text{mol}}}$$

$$= \frac{3.6225 \times 10^{12} \frac{\text{g}^2}{\text{mol}}}{5.85 \times 10^6 \text{g}} = 6.192 \times 10^5 \frac{\text{g}}{\text{mol}}$$

An easier way to calculate <Mw>is to use the following equation with calculated w_i values (1.0×10^5, 5.0×10^5, 2.0×10^6, 2.25×10^6, and 1.0×10^6 g):

$$<M_n> = \frac{\sum w_i M_i}{\sum w_i}$$

$$= \frac{\sum\left(1.0 \times 10^5 \times 1.0 \times 10^5 + 5.0 \times 10^5 \times 2.5 \times 10^5 + 2.0 \times 10^6 \times 4 \times 10^5 + 2.25 \times 10^6 \times 7.5 \times 10^5 + 1.0 \times 10^6 \times 1.0 \times 10^6\right)\frac{g^2}{mol}}{\sum\left(1.0 \times 10^5 + 5.0 \times 10^5 + 2.0 \times 10^6 + 2.25 \times 10^6 + 1.0 \times 10^6\right)g}$$

$$= \frac{3.6225 \times 10^{12}\frac{g^2}{mol}}{5.85 \times 10^6 g} = 6.192 \times 10^5 \frac{g}{mol}$$

There are different approaches to determine the molar mass of polymers/biopolymers—chemical (e.g., end-group analysis) and physical (based on colligative properties) [4, 25, 27]. To use the end-group analysis (determines $<M_n>$), the end groups should be quantifiable using standard analytical techniques. The measurement sensitivity drops significantly with the increase of chain length (molar mass) which decrease the concentration of end groups per unit volume [4].

Problem 4.3

To determine a biopolymer (linear with a monovalent acidic group at one end)'s molar mass, 250 μL of 1 M NaOH was used for titration. The biopolymer solution was prepared by dissolving 10 g of biopolymer powder into a 50 mL of water. Ignore the volume change during titration. What is this biopolymer's <Mn>?

Answer Since there is one acid group per biopolymer molecule, $n_{acidic\ group} = n_{biopolymer} = n_{NaOH}$.

$$\text{nNaOH} = 250\,\mu L \times 1M = 250 \times 10^{-6} L \times 1\text{mol/L} = 2.50 \times 10^{-4}\text{mol}.$$

$$<M_n> = w_{biopolymer}/n_{biopolymer} = 10g/\left(2.50 \times 10^{-4} mol\right) = 4 \times 10^{-4}\text{g/mol}.$$

Different colligative property measurements can be used to determine a polymer/biopolymer's molar mass, such as osmotic pressure, elevation of boiling point and depression of freezing point [27, 28]. For example, molar mass of a polymer/biopolymer can be determined by measuring its osmotic pressures (π) at different concentrations (c). The governing equations are:

$$\frac{\pi}{c} = RT\left(\frac{1}{M} + A_2c + A_3c^2 + \ldots\right) \tag{4.4}$$

$$\left(\frac{\pi}{c}\right)_{c\to 0} = \left(\frac{\pi}{c}\right)_0 = RT\left(\frac{1}{M} + 0 + 0 + \ldots\right) = \frac{RT}{M} \tag{4.5}$$

$$M = \frac{RT}{\left(\frac{\pi}{c}\right)_0} \tag{4.6}$$

$$\frac{\pi}{c} = \left(\frac{\pi}{c}\right)_0\left(1 + \Gamma_2c + \Gamma_3c^2 + \ldots\right) = \frac{RT}{M}\left(1 + \Gamma_2c + \Gamma_3c^2 + \ldots\right) \tag{4.7}$$

where R is the gas constant (8.314 J/(K × mol)), T is temperature in K. A_2, Γ_2 and A_2, Γ_3 are the second and third virial coefficients, respectively, describing the interaction between one polymer molecule and the solvent, and multiple polymer- solvent interactions. When solutions are sufficiently diluted (c is very low and the term Γ_3c^2can be neglected), a plot (π/c) against c is close to linear [27]. Molar mass can be calculated using Eq. 4.6 combined with the estimating $(\pi/c)_0$ using the plot (π/c) against c.

Problem 4.4*

The following osmotic pressures were measured for solutions of poly-isobutylene in benzene at 298 K. Results are reported as centimeters of solution, h, and need to be converted to pressures using the relationship: π = hdg, where d is the solution density and g the gravitational constant equal to 981 cm/s^2.

h (cm solution)	c (g/dm^3)
0.93	5.0
1.91	10.0
2.94	15.0
4.00	20.0

First, calculate the osmotic pressure using π = hdg (d = 880 kg/m^3) and π/c.

h (cm = 0.01 m)	Π (kg/(m × s^2) = J/m^3)	c (g/dm^3 = kg/m^3)	π/c (J/kg)
0.93	80.28504	5.0	16.057008
1.91	164.88648	10.0	16.488648
2.94	253.80432	15.0	16.920288
4.00	345.312	20.0	17.2656

Answer

Then plot (π/c) against c and then perform regression—the intercept is $(\pi/c)_0$.

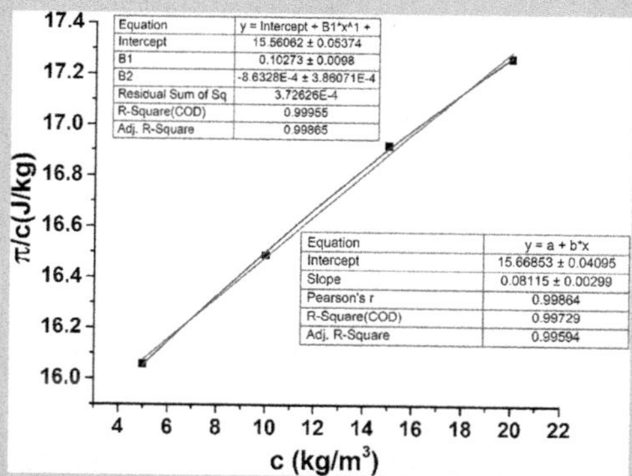

The results are close comparing quadratic (better fit based on the R^2 and theoretically) and linear regressions. Using $(\pi/c)_0 = 15.6$ J/kg,

$$M = \frac{RT}{\left(\frac{\pi}{c}\right)_0} = \frac{8.314\frac{J}{K\times mol}\times 298K}{15.6\frac{J}{kg}} = 158.8\frac{kg}{mol}.$$

*From [27], Copyright (© 2008) by CRC Press. Reproduced by permission of Taylor & Francis Group.

Molar mass can also be determined using relative methods, such as viscosity. It is logical to assume with other parameters being equal, that larger molecules would impede flow more than smaller ones, resulting in a higher solution viscosity. However, for relative methods, absolute methods need be employed for calibration (e.g., relating viscosity to molar mass) [4].

Polymer morphology is also an important factor to consider for characterization, which can significantly affect a polymer's thermal and mechanical properties. Polymer morphology is the overall form of polymer structure, including crystallinity, branching, molecular weight, cross-linking, and so on [29]. Based on the structure arrangements, we have crystalline (such as polyethylene), amorphous (such as polystyrene), and semicrystalline polymers. Amorphous polymers are generally found in a random coil conformation and have a disordered chain structure; while crystalline polymers are predominantly in the all-trans conformation and the chains are arranged in lamellae [30]. Crystalline polymers have a sharp melting point (at T_m) where they turn from a hard and solid material to a liquid; amorphous polymers have a softening range above the glass transition temperature (T_g)—transitions from a relatively brittle state into a viscous and rubbery state [31, 32]. As shown in Fig. 4.7, during melting, the polymer's temperature stays steady for a while even with heat input (melting requires energy and no energy goes into raising the polymer temperature). However, there is no latent heat of glass transition. Both melting and glass transition lead to the increase of heat capacity [33]. During melting, there is a transfer of heat between system and surroundings and the system undergoes an abrupt volume change (first-order transition) [34]. Semicrystalline polymers, containing both crystalline and amorphous structures, have both T_m and T_g. Most widely used polymers are semicrystalline [35]. The degree of crystallinity plays a key role in determining the type of morphology of semicrystalline structures. A decrease in crystallinity results in a morphology transition from spherulitic morphology (spherical crystal colonies) to individually crystalline lamellae randomly distributed in the amorphous matrix (Fig. 4.7) [36, 37]. The polymer chains in the amorphous regions predominantly tie one crystallite to another [38]. Unlike metal, semicrystalline polymers do not solidify (recrystallize) at its T_m. There remains a difference between its recrystallization temperature (T_r) and T_m [39].

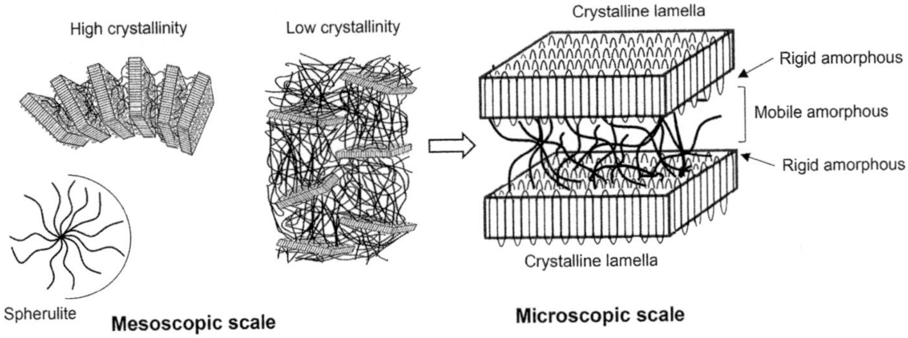

Fig. 4.7 Two-scale view of the semi-crystalline structure. *Adapted from* [37] *with permission from Elsevier*

Differential Scanning Calorimetry (DSC), a physical characterization method, is commonly used for thermal analysis of polymers/biopolymers. DSC involves supplying heat at a constant rate to the sample and a reference material to keep both at the same temperature. The heat flow necessary to supply difference is plotted against the temperature or time for obtaining a typical thermogram [40, 41]. The thermogram provides material properties, such as temperatures at the glass transition (T_g), onset of melting, peak of melting (T_m), and enthalpy of fusion (ΔH) for calculating crystallinity [42]. A DSC thermogram of a commercial poly(lactic acid) is shown in Fig. 4.8. For a specific sample, T_m is higher than T_g. Glass transition is better identified as a state transition rather than phase transition [40]. At T_g the segmental mobility of chains increases and a polymer is more elastic and flexible [41]. Some semi-crystalline polymers with slow crystallization ability, such as poly(lactic acid), do not have time to crystallize during cooling and thus crystallize during a 2nd heating cycle (cold crystallization) [41]. It is also important to note that the heating rate can greatly affect the glass transition and recrystallization temperatures [43]. The cross-links between chains restrict rotational motion and raise the T_g. Hence, higher cross-linking shows higher T_g [44]. Between T_g and T_m, crystalline region remain unaffected—rigid crystallites dispersed in rubbery amorphous matrix [44].

Regarding the mechanical properties, an engineering stress–strain curve is commonly used. It is determined by measuring elongation as a function of continuously increasing loads of standardized test specimens (dog bone-shaped). Engineering stress is the applied

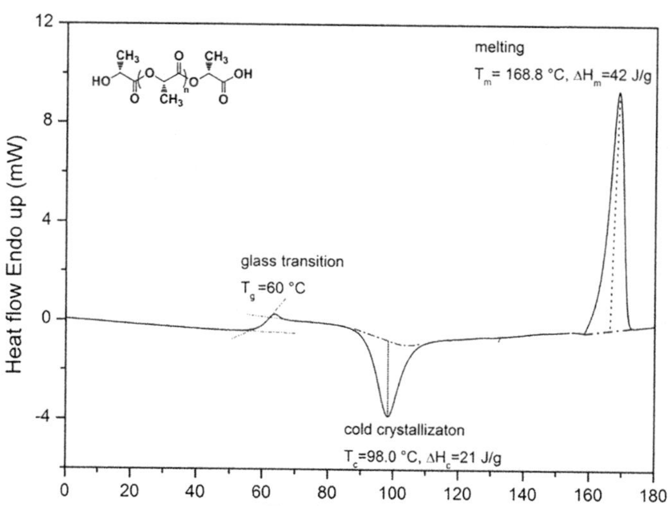

Fig. 4.8 DSC thermogram of a commercial poly(lactic acid) product (molar mass of 70,400 g/mol and PDI = 1.8) detected during 2nd heating cycle (0–180 °C, 10 °C/min, N_2 atmosphere). *Adapted from* [41]

load (axial force) divided by the original cross-section area of a material (the actual cross-section area of a material changes during testing). Strain is defined as the change in length divided by the original length. At low load levels, stress varies linearly (reflected by the Young's modulus) with the strain in the elastic regime (deformation/response is reversible). With increased loading, the response curve slope decreases, but remains linear. Then, the deformation becomes plastic (irreversible). Yield strength is the minimum stress under which a material deforms permanently. It is difficult to determine the stress at which the plastic deformation starts. Hence, yield strengths are determined at some level of plastic deformation, typically 0.2%, to establish offset yield strength. It is determined by finding the intersection of the stress–strain curve with a line (the strain is 0.002) parallel to the initial slope (linear region) of the curve. Ultimate tensile strength is the maximum stress achieved in the test (necking starts) [45, 46]. A typical engineering stress–strain curve is shown in Fig. 4.9. The stress keeps increasing until the material breaks, even after the ultimate tensile strength, for a stress–strain curve. Microstructures of a semicrystalline polymer at various stages are shown in Fig. 4.10. The strength of a polymer/biopolymer mainly depends on the degree of crystallinity, temperature, and density among others. The mechanical response of the crystalline region is elastic while the response in the amorphous region shows both elastic and viscoelastic behaviors. As the stress increases, the load carried dominantly by the crystallites shift to the amorphous regions until the chains are fully disentangled [38].

Nuclear magnetic resonance spectroscopy (NMR) and Fourier transform infrared spectroscopy (FT-IR) are powerful tools for studying polymer/biopolymer chemistry. NMR provides structure information, while FTIR is useful for identifying functional groups. They are complementary. The guiding principle behind NMR is that some nuclei exist in specific nuclear spin states when exposed to an external magnetic field. NMR detects transitions between these spin states that are specific to the particular nuclei in question and the nuclei's chemical environment. NMR generally detects one isotope at a time (such

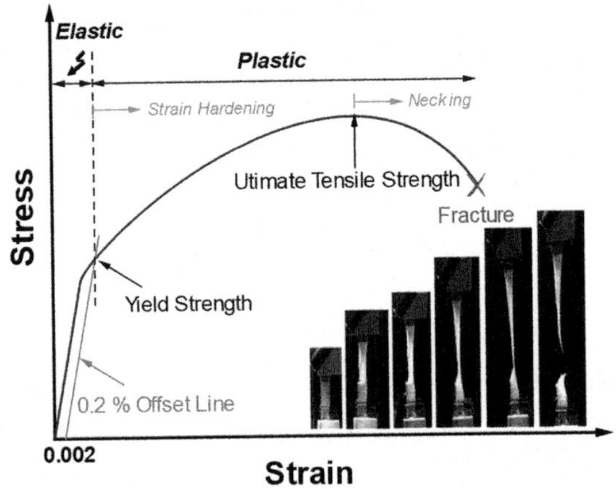

Fig. 4.9 A typical engineering stress–strain curve. The insert is adapted from [38]

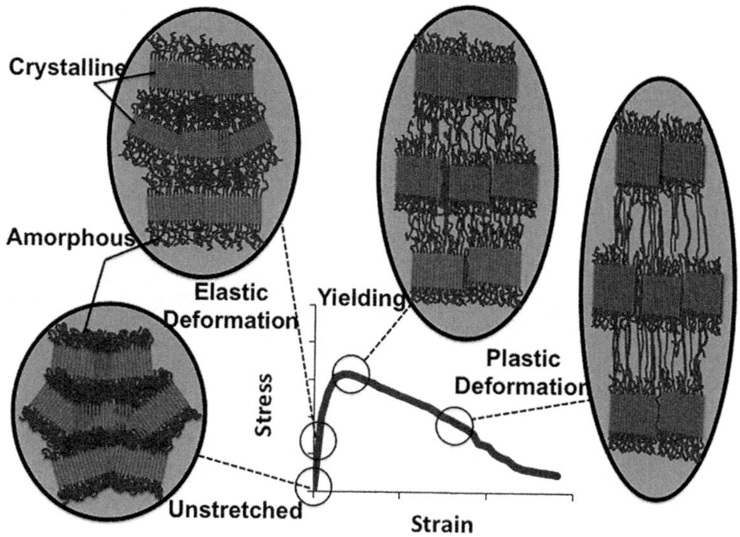

Fig. 4.10 Schematic of semicrystalline polymer microstructure at various stages of tensile deformation. *Adapted from* [38]

as ^{1}H NMR and ^{13}C NMR)—differentiating between different elements and isotopes due to the fact that each specific nuclei only absorb at a specific frequency. Since nuclei of the same type in different electron environments will be shielded at different levels, the difference in their environment is defined as the chemical shift (δ) and expressed in parts per million (ppm). NMR spectroscopy works by varying the machine's emitted frequency over a small range [47, 48]. Using NMR, chemical shift (chemical shielding) is commonly used for the structural characterization of polymers. The chemical shifts of atoms in a polymer chain or multiple polymer chains depends on the electronic and molecular environments of the nuclei [49]. The signals can be assigned to specific atoms along the polymer backbone and side chains [50]. In the solution state, the NMR chemical shifts of polymers are generally the averaged values over all the possible conformations. In solids, the chemical shifts are often characteristic specific conformations due to the strongly restricted rotation around the bonds [49]. In general, protons bonded to more substituted carbon atoms resonate at higher ppm. The stronger the electron-withdrawing group, the higher their ppm value (the adjacent protons are more de-shielded). One common application of NMR in biopolymer engineering is to confirm the successful modification of a biopolymer. For example, ^{1}H-NMR analysis was performed to confirm the successful methacrylation of both chitosan and chondroitin sulfate (Fig. 4.11) [51]. NMR has also been used for molar mass analysis of polymers/biopolymers [52, 53].

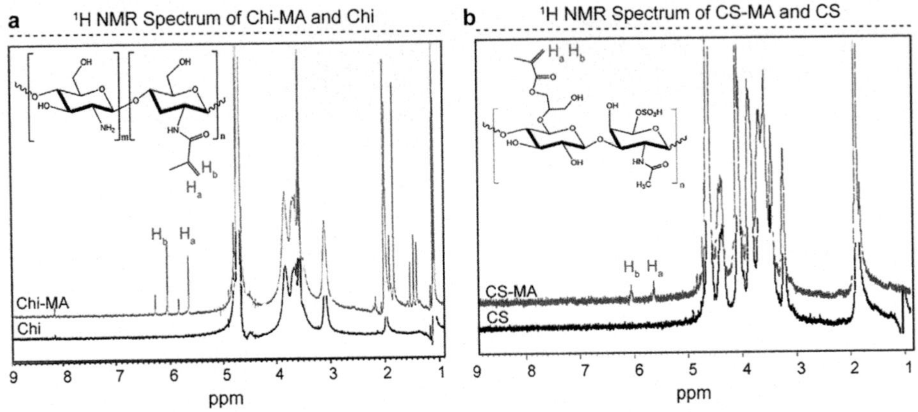

Fig. 4.11 ^{1}H-NMR spectra of **a** Chi-MA and **b** CS-MA. The peaks corresponding to the methacrylate group were confirmed (H_a and H_b) with respective molecular structures, which indicated that both Chi-MA and CS-MA were successfully synthesized. Chi: chitosan, CS: chondroitin sulfate. *Adapted from* [51]*with permission from John Wiley and Sons, Inc*

Regarding FT-IR (commonly in the region of 4000–400 cm^{-1}), IR light is absorbed by the functional groups with strong dipoles and display strong peaks by selectively absorbing radiation of specific wavelengths [54, 55]. Infrared radiation is the proper energy to excite vibrations in molecules. Microwave radiation causes rotation, while X-rays cause bond breaking and ionization. The number of vibrations is determined by the degrees of freedom of the molecule (3N-6 for non-linear molecules and N is number of atoms) [56–58]. As shown in Fig. 4.12, the two fundamental vibrations are stretching (distance between the atoms and the bond length are affected) and bending (slant between the bonds is altered) vibration. Spectra can be analyzed looking at two major regions, the functional group (4000–1300 cm^{-1}) and the fingerprint (1300–600 cm^{-1}) regions. The functional group region can be further subdivided into specific zones representative of different functional groups. The fingerprint region is specific to the molecular structure of the sample, which can be used for authentication [59]. A common application of FTIR is to compare the biopolymers/biopolymer systems before and after specific modifications. For example, FTIR was used to compare the Ca-pectin microspheres (PM) before and after crosslinking—EDC-crosslinked (EPM) and transglutaminase-crosslinked (TGM) Ca-pectin microspheres. As shown in Fig. 4.13, in the PM spectrum, the band at 1592 cm^{-1} corresponds to carbonyl stretching of COO− groups. Around 1634 cm^{-1} of the EGM spectrum is attributed to the COO− of pectin, amide I bond regions of gelatin, and amide bond formation, while the peak at 1541 cm^{-1} represents the amide II region due to the gelatin coating. The TGM spectrum shows results similar to the EGM spectrum considering the only difference was the crosslinking mechanism of pectin and gelatin [60].

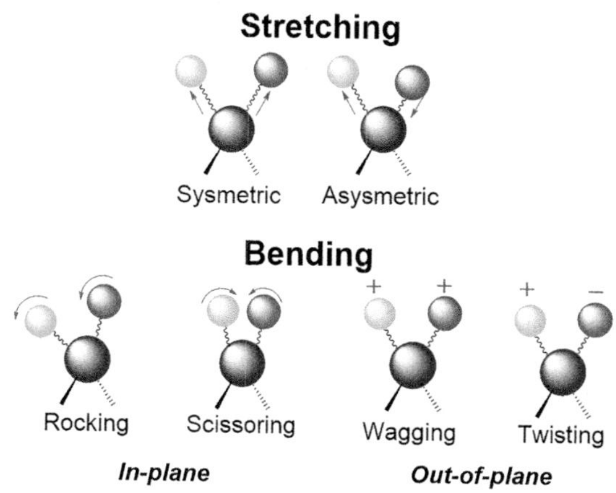

Fig. 4.12 Different vibrational models. Rocking: both atoms moving in -plane either clockwise or anticlockwise, Scissoring: both atoms simultaneously moving in-plane either toward each other or away from each other, Wagging: both atoms simultaneously move like a V sign back and forth, Twisting: one atom moves forward while the other moves backward. *Generated based on* [58, 59]

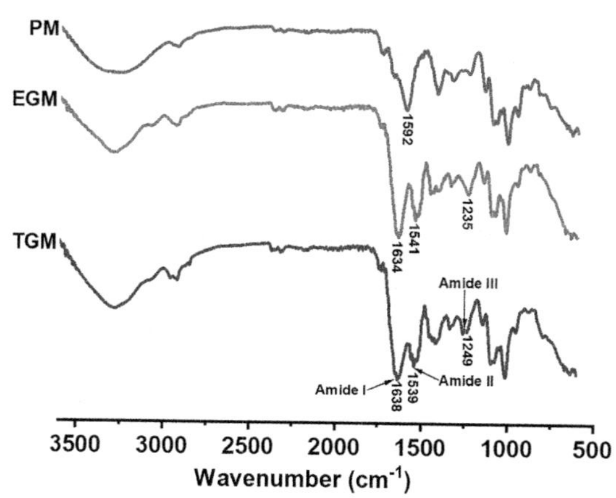

Fig. 4.13 ATR-FTIR spectra of Ca-pectin (GM), EDC-crosslinked GM (EGM), and transglutaminase-crosslinked GM (TGM) microspheres. ATR: attenuated total reflection, EDC: 1-ethyl-3-(3-dimethylaminopropyl) carbodiimide hydrochloride. *Adapted from* [60]

For biomedical applications, biocompatibility is critical. Biocompatibility is defined as the ability of a material to perform with an appropriate host response in a specific situation [61]. The definition has been evolving with time, especially considering biomaterials are being used in increasingly diverse and complex situations as well as new applications [62]. A newer definition of biocompatibility is the ability of a material to perform their desired functions with respect to a medical therapy, to induce an appropriate host response in a specific application and to interact with living systems without having any risk of injury, toxicity, or rejection by the immune system and undesirable or inappropriate local or systemic effects [63]. When considering biocompatibility, the characteristics of both the material and the biological host within which the material is placed should be considered. In fact, biocompatibility is a characteristic of a material-biological host system rather than a property of a material. Take polytetrafluoroethylene (PTFE) for example, monolithic PTFE products tested and used clinically have been found to pass all preclinical biological safety tests and are generally considered biocompatible. However, for situations such as cell adhesion, it is not compatible as PTFE is not preferred for cell attachment/adhesion [64].

Regarding biocompatibility testing, there are various standards from different organizations, such as International Standards Organization (ISO) and American Society for Testing and Materials (ASTM) (Table 4.2). Take cytotoxicity for example, relevant standards include ISO 10993–5:2009, ASTM F813-07, and ASTM F895-84 [63, 65]. During biocompatibility evaluation, the factors that should be considered include: nature of contact, type of contact, frequency and duration of contact, and materials (compositions of a product) [66]. In vitro testing should be conducted first before moving to animal models. Cytotoxicity testing determines the effects of materials on living cells. The primary endpoints include cell viability, morphological changes, cell detachment, cell lysis, and cell proliferation and differentiation (for certain applications) [67]. If the materials will come into contact with blood, hemocompatibility should be assessed. In vitro hemocompatibility of a material can be evaluated through coagulation and hemolysis and such [68]. For animal studies, the proposed animal research protocol has to be reviewed by an Institutional Animal Care and Use committee (IACUC). The IACUC is responsible for oversight of the animal care and use program—protocol review, postapproval monitoring and more [69]. For cytotoxicity testing, various factors can affect the results, such as the assay chosen for evaluating cell viability. A recent study showed that the nanofibrous (polymeric) materials affect the reaction of cytotoxicity assays (MTT, XTT, CCK-8, alamaBlue and PrestoBlue). As shown in Fig. 4.14, according to the ISO EN 10,993–5 protocol's criteria (70% threshold), the assays showed mixed results [70].

For certain applications, degradability should be also be assessed. The degradation of polymers and biopolymers generally refers to complex phenomena that, under environmental conditions, lead to changes in chemical composition, e.g. appearance, disappearance, transformation of chemical groups, and changes in physical properties and performance, e.g. loss of mechanical properties, brightness, discoloration, etc. [71]. Degradation of polymers/biopolymers can involve different mechanisms either simultaneously or sequentially—thermal degradation, mechanical degradation (from mechanical

Table 4.2 International standards for biocompatibility testing of biomaterials adapted from [65] with permission from Spandidos Publications.

Biological response	Standards[a]	
	ISO	ASTM
Cytotoxicity	10,993–5	F813-07; F895-84; F1027-06
Sensitization	10,993–10	F720-81; F2147-01; F2148-07
Irritation	10,993–10	F719-81; F749-98
Acute systemic toxicity	10,993–11	F750-87
Subacute toxicity	10,993–11	–
Genotoxicity	10,993–3	E1262-88
Immunoesponsiveness	10,993–20	F1906-98
Hemocompatibility	10,993–4	F756-08
Chronic toxicity	10,993–11	–
Carcinogenicity	10,993–3	F1439-03
Degradation	10,993–9; 10,993–13	F1983-14
	10,993–14; 10,993–15	
Implantation	10,993–6	F1408-97; F763-04;
		F1904-98
		F981-04; F1983-99

[a]ISO, International Standards Organizations; ATSM, American Society for Testing and Materials

stress), photodegradation (from the exposure to sunlight and such), oxidative degradation (from oxidation), and hydrolytic degradation (from hydrolysis) [71, 72]. Besides the environmetal factors, the chemical composition and structure of a polymer/biopolymer signnifcantly affects its degradation: molar mass, chemical bonds and funtional groups contained, morphology (degree of crystallinity/extend surface area). For example, polymers/biopolymers that only contain C–C bonds in their backbone are generally resistant to degradation, while polymers/biopolymers with ester bonds are more prone to degradation. Moreover, polymers/biopolymers with low crystallinity and high surface area tend to be degradable [72]. Another term that needs attention related to degradation is biodegradation, which refers to a collection of biological processes that breakdown and convert organic matter to inorganic matter (carriered out by microoranisms or other biological agents) [73, 74]. It is recommended that degradation process takes place in three stages: biodeterioration (or bio biofilm formation; the change of macro materials into small fractions under the cooperative action of different microorganisms and/or abiotic agents), depolymerization (or biofragmentation; enzyme (microbial extracellular)-catalyzed reduction of molar mass of the polymers), and bio-assimilation (small molecules produced integrate with the microbial metabolism) or mineralization (occurs concurrently with

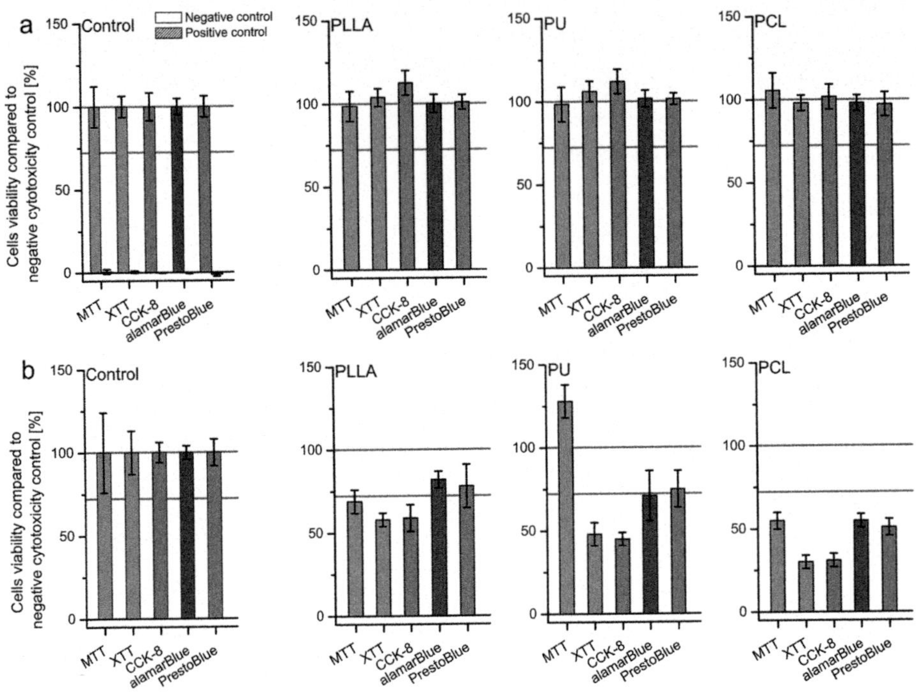

Fig. 4.14 Cells viability was tested by MTT, XTT, CCK-8, alamarBlue, and PrestoBlue assays **a** on scaffolds' liquid extracts after 24 h of incubation with PLLA, PU, and PCL; and **b** after 24 h of direct culture on PLLA, PU, and PCL scaffolds; including controls. In every graph, the green line represents the negative control result (100%), and a red line represents the threshold for cytotoxicity level of 70% of cells viability compared to the negative control. PLLA: poly-L-lactic acid, PU: polyurehane, PCL: polycaprolactone (PCL). *Adapted from* [70]

assimilation; organic materials are converted into minerals) [75–77]. Biodegradation can be either aerobic or anerobic which leads to different products – biomass, H_2O and CO_2 (aerobic) and biomass, H_2O, CO_2 and CH_4 (anaerobic). Like degradation, biodegradation is affected by polymer/biopolymer morphology, structure, chemical treatement and molar mass [77] (Fig. 4.15).

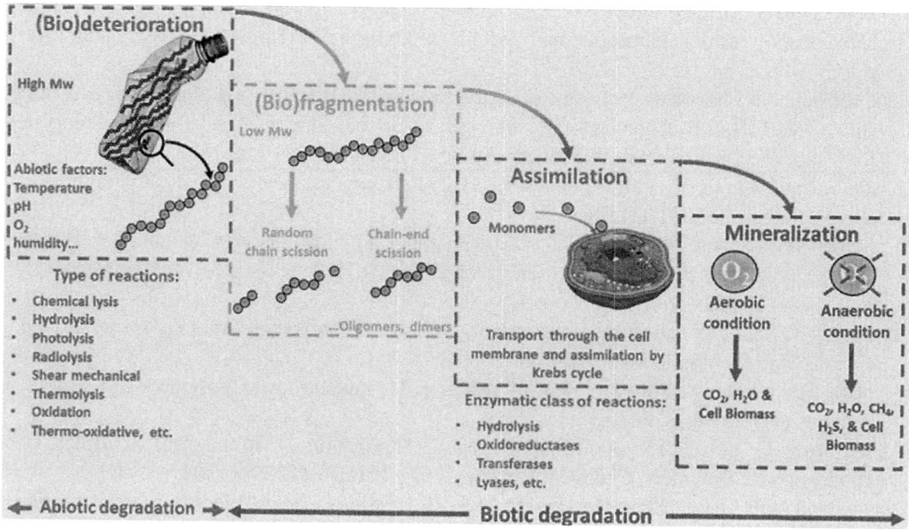

Fig. 4.15 Steps of polymer degradation. *Adapted from* [76]

References

1. Alves A, Lasmar D, Miranda I, et al. The Potential of Activated Carbon in the Treatment of Water for Human Consumption, a Study of the State of the Art and Its Techniques Used for Its Development. Advances in Bioscience and Biotechnology. 2021;12:143–153.
2. Omar BA, Elmasry R, Eita A, et al. Upgrading the preparation of high-quality chitosan from Procambarus clarkii wastes over the traditional isolation of shrimp chitosan. Saudi Journal of Biological Sciences. 2022;29(2):911–919.
3. Mohapatra S. Chapter 59 - Sterilization and Disinfection. In: Prabhakar H, editor. Essentials of Neuroanesthesia: Academic Press; 2017. p. 929–944.
4. Brazel CS, Rosen SL. Fundamental Principles of Polymeric Materials. Wiley; 2012.
5. Farag H, Sisinni E, Gidlund M, et al. Priority-Aware Wireless Fieldbus Protocol for Mixed-Criticality Industrial Wireless Sensor Networks. IEEE Sensors Journal. 2019;19(7):2767–2780.
6. Francis LF. Chapter 3—Melt Processes. In: Francis LF, editor. Materials Processing. Boston: Academic Press; 2016. p. 105–249.
7. Svečko R, Kusić D, Kek T, et al. Acoustic Emission Detection of Macro-Cracks on Engraving Tool Steel Inserts during the Injection Molding Cycle Using PZT Sensors. Sensors. 2013;13(5):6365–6379.
8. Zhao N-y, Lian J-y, Wang P-f, et al. Recent progress in minimizing the warpage and shrinkage deformations by the optimization of process parameters in plastic injection molding: a review. The International Journal of Advanced Manufacturing Technology. 2022;120(1):85–101.
9. Belcher SL. 16 - Blow Molding. In: Kutz M, editor. Applied Plastics Engineering Handbook. Oxford: William Andrew Publishing; 2011. p. 267–288.

10. Ebnesajjad S, Morgan RA. 11—Applications of Processing Aid Additives. In: Ebnesajjad S, Morgan RA, editors. Fluoropolymer Additives. Oxford: William Andrew Publishing; 2012. p. 193–209.
11. Cherrington R, Goodship V. 1—Introduction to Multifunctionality and Manufacture. In: Goodship V, Middleton B, Cherrington R, editors. Design and Manufacture of Plastic Components for Multifunctionality. Oxford: William Andrew Publishing; 2016. p. 1–18.
12. Jin Y, Lin J, Cheng Y, et al. Lignin-Based High-Performance Fibers by Textile Spinning Techniques. Materials. 2021;14(12):3378.
13. Mahat MM, Sabere ASM, Azizi J, et al. Potential Applications of Conducting Polymers to Reduce Secondary Bacterial Infections among COVID-19 Patients: a Review. Emergent Materials. 2021;4(1):279–292.
14. Zhang W. Nanofabrication I: Electrospinning of Nanofibers. Nanotechnology for Bioengineers. Cham: Springer International Publishing; 2020. p. 7–13.
15. Nath SD, Nilufar S. An Overview of Additive Manufacturing of Polymers and Associated Composites. Polymers. 2020;12(11):2719.
16. Kirillova A, Yeazel TR, Asheghali D, et al. Fabrication of Biomedical Scaffolds Using Biodegradable Polymers. Chemical Reviews. 2021;121(18):11238–11304.
17. Mobarak MH, Islam MA, Hossain N, et al. Recent advances of additive manufacturing in implant fabrication – A review. Applied Surface Science Advances. 2023;18:100462.
18. Saleh Alghamdi S, John S, Roy Choudhury N, et al. Additive Manufacturing of Polymer Materials: Progress, Promise and Challenges. Polymers. 2021;13(5):753.
19. Islam MA, Mobarak MH, Rimon MIH, et al. Additive manufacturing in polymer research: Advances, synthesis, and applications. Polymer Testing. 2024;132:108364.
20. Jafferson JM, Chatterjee D. A review on polymeric materials in additive manufacturing. Materials Today: Proceedings. 2021;46:1349–1365.
21. Chyr G, DeSimone JM. Review of high-performance sustainable polymers in additive manufacturing. Green Chemistry. 2023;25(2):453–466.
22. Zhang W, Chen J, Zeng H. Chapter 8 - Polymer processing and rheology. In: Narain R, editor. Polymer Science and Nanotechnology: Elsevier; 2020. p. 149–178.
23. Foroughi-Dahr M, Mostoufi N, Sotudeh-Gharebagh R, et al. Particle Coating in Fluidized Beds. Reference Module in Chemistry, Molecular Sciences and Chemical Engineering: Elsevier; 2017.
24. Liu Y, Hao M, Chen Z, et al. A review on recent advances in application of electrospun nanofiber materials as biosensors. Current Opinion in Biomedical Engineering. 2020;13:174–189.
25. Carraher CE. Introduction to Polymer Chemistry. CRC Press; 2017.
26. Shrivastava A. 2 - Polymerization. In: Shrivastava A, editor. Introduction to Plastics Engineering: William Andrew Publishing; 2018. p. 17–48.
27. Cowie JMG, Arrighi V. Polymers: Chemistry and Physics of Modern Materials, Third Edition. Informa Healthcare; 2007.
28. Chanda M. Introduction to Polymer Science and Chemistry: A Problem-Solving Approach, Second Edition. Taylor & Francis; 2013.
29. Whisnant D. Polymer Chemistry: Morphology. 2021. Available from: https://eng.libretexts.org/@go/page/2909
30. Crystalline and Amorphous Polymers. 2023. Available from: https://eng.libretexts.org/@go/page/7865
31. Goodship V. 4—Injection Molding of Thermoplastics. In: Goodship V, Middleton B, Cherrington R, editors. Design and Manufacture of Plastic Components for Multifunctionality. Oxford: William Andrew Publishing; 2016. p. 103–170.

32. Khalifeh M, Saasen A. Types of Permanent Plugging Materials. Introduction to Permanent Plug and Abandonment of Wells. Cham: Springer International Publishing; 2020. p. 97–136.
33. The Glass Transition. Polymer Science Learing Center. Available from: https://pslc.ws/macrog/tg.htm
34. Whisnant D. Polymer Chemistry: The Glass Transition. 2022. Available from: https://eng.libretexts.org/@go/page/2924
35. Sarath CC, Shanks RA, Thomas S. Chapter 1 - Polymer Blends. In: Thomas S, Shanks R, Chandrasekharakurup S, editors. Nanostructured Polymer Blends. Oxford: William Andrew Publishing; 2014. p. 1–14.
36. Mirkhalaf M, Vadizadeh R. Micro-mechanical modeling of semi-crystalline polymers: A review. International Journal of Solids and Structures. 2024;290:112691.
37. Yan Z, Guo Q, Zaïri F, et al. Continuum-based modeling large-strain plastic deformation of semi-crystalline polyethylene systems: Implication of texturing and amorphicity. Mechanics of Materials. 2021;162:104060.
38. Kelly T, Ghadi BM, Berg S, et al. In Situ Study of Strain-Dependent Ion Conductivity of Stretchable Polyethylene Oxide Electrolyte. Scientific Reports. 2016;6(1):20128.
39. Kumar S. 10.05 - Selective Laser Sintering/Melting. In: Hashmi S, Batalha GF, Van Tyne CJ, et al., editors. Comprehensive Materials Processing. Oxford: Elsevier; 2014. p. 93–134.
40. Leyva-Porras C, Cruz-Alcantar P, Espinosa-Solís V, et al. Application of Differential Scanning Calorimetry (DSC) and Modulated Differential Scanning Calorimetry (MDSC) in Food and Drug Industries. Polymers. 2020;12(1):5.
41. Gregorova A. Application of Differential Scanning Calorimetry to the Characterization of Biopolymers. In: Elkordy AA, editor. Applications of Calorimetry in a Wide Context - Differential Scanning Calorimetry, Iso thermal Titration Calorimetry and Microcalorimetry. IntechOpen; 2013.
42. Lynch JM, Corniuk RN, Brignac KC, et al. Differential scanning calorimetry (DSC): An important tool for polymer identification and characterization of plastic marine debris. Environmental Pollution. 2024;346:123607.
43. Wellen RMR, Canedo E, Rabello MS. Nonisothermal cold crystallization of poly(ethylene terephthalate). Journal of Materials Research. 2011;26(9):1107–1115.
44. Physical, Thermal, and Mechanical Properties of Polymers. Biosurfaces2014. p. 329–344.
45. Horwood A, Chockalingam N. Chapter 2 - Principles of materials science. In: Horwood A, Chockalingam N, editors. Clinical Biomechanics in Human Locomotion: Academic Press; 2023. p. 91–174.
46. Krauss G. Strengthening Mechanisms in Steels. In: Buschow KHJ, Cahn RW, Flemings MC, et al., editors. Encyclopedia of Materials: Science and Technology. Oxford: Elsevier; 2001. p. 8870–8881.
47. Raja PMV, Barron AR. NMR Spectroscopy. 2022. Available from: https://chem.libretexts.org/@go/page/55887
48. Chemical Shifts in ^{1}H NMR Spectroscopy. 2024. Available from: https://chem.libretexts.org/@go/page/31540
49. Ando I. Some aspects of the NMR chemical shift/structure correlation in the structural characterization of polymers and biopolymers. Polymer Journal. 2012;44(8):734–747.
50. Mirau PA. NMR characterisation of polymers. In: Hunt BJ, James MI, editors. Polymer Characterisation. Dordrecht: Springer Netherlands; 1993. p. 37–68.
51. Park MJ, An Y-H, Choi YH, et al. Enhanced Neovascularization Using Injectable and rhVEGF-Releasing Cryogel Microparticles. Macromolecular Bioscience. 2021;21(11):2100234.
52. Izunobi JU, Higginbotham CL. Polymer Molecular Weight Analysis by 1H NMR Spectroscopy. Journal of Chemical Education. 2011;88(8):1098–1104.

53. Harrell ML, Bergbreiter DE. Using 1H NMR Spectra of Polymers and Polymer Products To Illustrate Concepts in Organic Chemistry. Journal of Chemical Education. 2017 2017;94(11):1668–1673.
54. Sharma A, Saini AK, Kumar N, et al. Methods of preparation of metal-doped and hybrid tungsten oxide nanoparticles for anticancer, antibacterial, and biosensing applications. Surfaces and Interfaces. 2022;28:101641.
55. How an FTIR Spectrometer Operates. 2023. Available from: https://chem.libretexts.org/@go/page/1844
56. Introduction to Vibrations. 2023. Available from: https://chem.libretexts.org/@go/page/77584
57. Wenzel T. Introduction to Infrared Spectroscopy. 2023. Available from: https://chem.libretexts.org/@go/page/111678
58. El-Azazy M. Introductory Chapter: Infrared Spectroscopy—A Synopsis of the Fundamentals and Applications. In: El-Azazy M, editor. Infrared Spectroscopy—Principles, Advances, and Applications. IntechOpen; 2018.
59. Mendes E, Duarte N. Mid-Infrared Spectroscopy as a Valuable Tool to Tackle Food Analysis: A Literature Review on Coffee, Dairies, Honey, Olive Oil and Wine. Foods. 2021;10(2):477.
60. Zieman J, Cohan M, Wang Y, et al. Development of Gelatin-Coated Hydrogel Microspheres for Novel Bioink Design: A Crosslinker Study. Pharmaceutics. 2023;15(1):90.
61. Williams DF, Biomaterials ESf. Definitions in Biomaterials: Proceedings of a Consensus Conference of the European Society for Biomaterials, Chester, England, March 3–5, 1986. Elsevier; 1987. (v. 4).
62. Williams DF. On the mechanisms of biocompatibility. Biomaterials. 2008;29(20):2941–2953.
63. Ghasemi-Mobarakeh L, Kolahreez D, Ramakrishna S, et al. Key terminology in biomaterials and biocompatibility. Current Opinion in Biomedical Engineering. 2019;10:45–50.
64. Williams DF. There is no such thing as a biocompatible material. Biomaterials. 2014;35(38):10009–10014.
65. Huzum B, Puha B, Necoara RM, et al. Biocompatibility assessment of biomaterials used in orthopedic devices: An overview. Exp Ther Med. 2021;22(5):1315.
66. U.S. Food and Drug Administration. Basics of Biocompatibility: Information Needed for Assessment by the FDA. Available from: https://www.fda.gov/medical-devices/biocompatibility-assessment-resource-center/basics-biocompatibility-information-needed-assessment-fda
67. Kandárová H, Pôbiš P. The "Big Three" in biocompatibility testing of medical devices: implementation of alternatives to animal experimentation-are we there yet? Front Toxicol. 2023;5:1337468.
68. Nalezinková M. In vitro hemocompatibility testing of medical devices. Thrombosis Research. 2020;195:146–150.
69. The IACUC. National Institutes of Health Office of Laboratory Animal Welfare. Available from: https://olaw.nih.gov/resources/tutorial/iacuc.htm
70. Podgórski R, Wojasiński M, Ciach T. Nanofibrous materials affect the reaction of cytotoxicity assays. Scientific Reports. 2022;12(1):9047.
71. Dintcheva NT. Overview of polymers and biopolymers degradation and stabilization towards sustainability and materials circularity. Polymer. 2024;306:127136.
72. Stevens ES. Green Plastics: An Introduction to the New Science of Biodegradable Plastics. Princeton University Press; 2002.
73. Mastropetros SG, Pispas K, Zagklis D, et al. Biopolymers production from microalgae and cyanobacteria cultivated in wastewater: Recent advances. Biotechnology Advances. 2022;60:107999.
74. Wu Y, Li T, Yang L. Mechanisms of removing pollutants from aqueous solutions by microorganisms and their aggregates: A review. Bioresource Technology. 2012;107:10–18.

75. Nizamuddin S, Baloch AJ, Chen C, et al. Bio-based plastics, biodegradable plastics, and compostable plastics: biodegradation mechanism, biodegradability standards and environmental stratagem. International Biodeterioration & Biodegradation. 2024;195:105887.
76. Silva RRA, Marques CS, Arruda TR, et al. Biodegradation of Polymers: Stages, Measurement, Standards and Prospects. Macromol. 2023;3(2):371–399.
77. Samir A, Ashour FH, Hakim AAA, et al. Recent advances in biodegradable polymers for sustainable applications. npj Materials Degradation. 2022;6(1):68.

Biopolymers: Biomedical Applications 5

5.1 Introduction to Biomedical Applications of Biopolymers

Biopolymers have been applied in various fields of biomedical engineering: medical devices, biosensing, surgical sutures/sealants, wound healing, drug delivery, and cellular and tissue engineering (Fig. 5.1) [1–6]. Generally, biopolymers are not used for bearing loads (such as body weight), compared to metals and ceramics, for biomedical applications [7]. Biodegradable polymers are generally preferred for applications inside the human body. This chapters focuses on drug delivery and cellular and tissue engineering applications of biopolymers.

Both synthetic (such as PLA (poly(lactic acid)), PGA (poly(glycolic acid)), PLGA (poly(lactic-*co*-glycolic) acid), PCL (polycaprolactone), and natural (such as alginate, pectin, chitosan, and collagen/gelatin) biopolymers have been used for biomedical applications. Take PLGA as an example, PLGA is one of the widely-used synthetic biodegradable biopolymers in biomedical applications. It is a copolymer of lactic acid and glycolic acid. The ratio of lactic acid and glycolic acid affects the polymer's properties such as degradation rate – the rate decreases with the increase of the ratio [8, 9]. Synthesized PLGA has two different structures: acid-terminated (PLGA–COOH) and ester-terminated (PLGA–$COOCH_3$). Ester-terminated PLGA is more hydrophobic and degrades slower [10]. Interestingly, the end groups can significantly affect drug encapsulation efficiency and loading capacity when it comes to drug delivery applications [11, 12] (Fig. 5.2).

Regarding natural biopolymers, polysaccharides such as alginate, pectin and chitin/chitosan have been discussed previously. Chitosan, in particular, is the second most commonly used biomaterial in pharmaceutical and biomedical fields, which has been used for drug delivery, tissue engineering, biosensing, bioimaging, and wound dressing [13]. There are various chitosan-based products in market, such as Tricol Biomedical's HemoCon® Patch. Chitosan is positively-charged, attracting negatively-charged red blood cells and

W. Zhang, *Biopolymer Engineering*, Synthesis Lectures on Chemical Engineering and Biochemical Engineering, https://doi.org/10.1007/978-3-031-98868-4_5

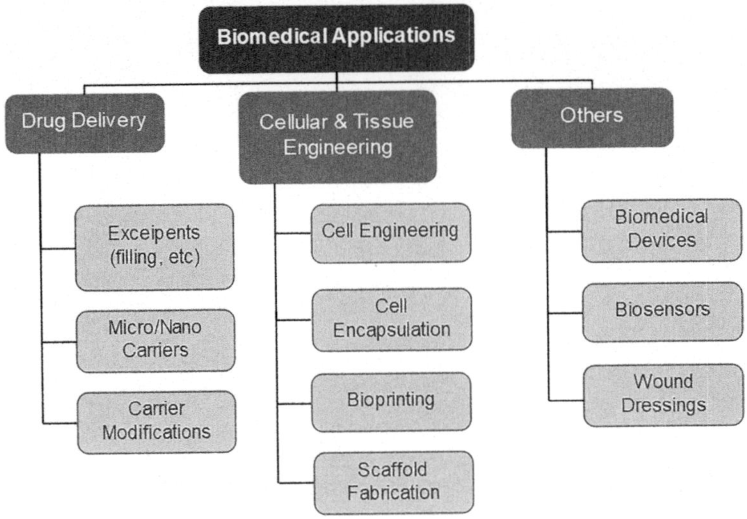

Fig. 5.1 Biomedical applications of biopolymers

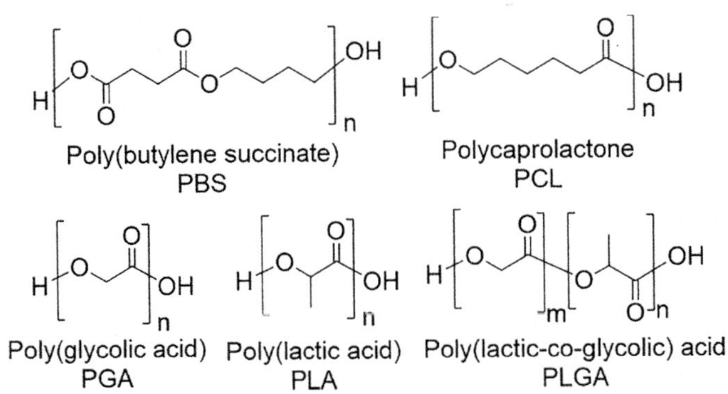

Fig. 5.2 Synthetic biodegradable polymers in biomedical applications

platelets. This interaction creates a supportive, primary seal at the wound site. Moreover, it also provides an antibacterial barrier [14]. Collagen is the most abundant structural protein in all animals (~ 30% of the total protein content). There are about 28 types of collagen that have been identified/described, with type I as the most common. All types of collagen contain three identical or different α-chains and at least one triple-helix domain [15]. Collagen molecules can be self-assembled into highly diverse shapes with intact triple-helical structures under certain conditions [16]. The biosynthesis of collagen molecules and self-assembly of collagen into fibers are shown in Fig. 5.3. The amino acid sequence indicates every third residue is glycine (Gly), resulting in a Gly-X-Y repeating sequence (X and Y are often proline (Pro) and hydroxyproline (Hyp), respectively) [17]. Glycine, without a side chain, allows the collagen chains to become close to each other [18]. Hydrogen bonds stabilize the collagen's triple-helix structure (between glycine N–H and proline C=O). And in some cases, the water molecule is involved in hydrogen bond formation (Fig. 5.4) [17, 19, 20]. Gelatin is produced by the partial hydrolysis of collagen. Depending on the pre-treatment procedures, there are two types of gelatin—type A (pI ≈ 8–9) and B (pI ≈ 4–5). Unlike collagen, gelatin solutions can gel (peptide chains form partially ordered helical structures (coil to helix)) upon cooling (~ < 30 °C). The process is thermal-reversible [15, 16].

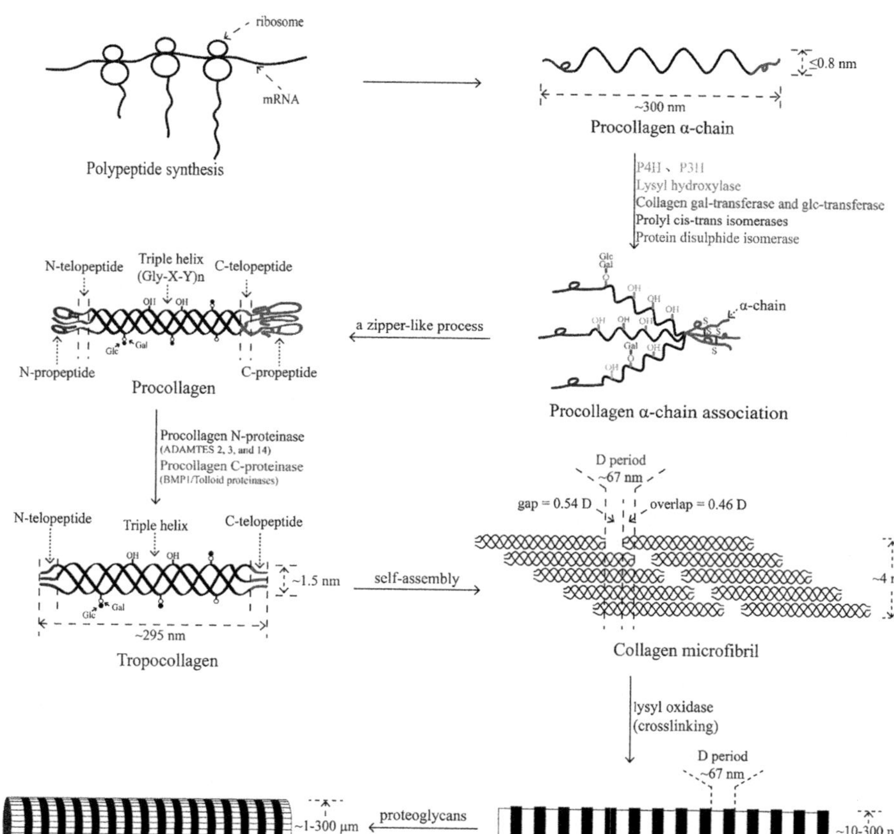

Fig. 5.3 Schematic of the biosynthesis process of collagen and self-assembly of collagen into fibers in vivo. P4H: prolyl 4-hydroxylase, P3H: prolyl 3-hydroxylase, ADAMTES: a-disintegrin-and-metalloproteinase-with-thrombospondin-like-motifs family, BMP1: bone morphogenetic protein 1. *Adapted from* [15] *with permission from Elsevier*

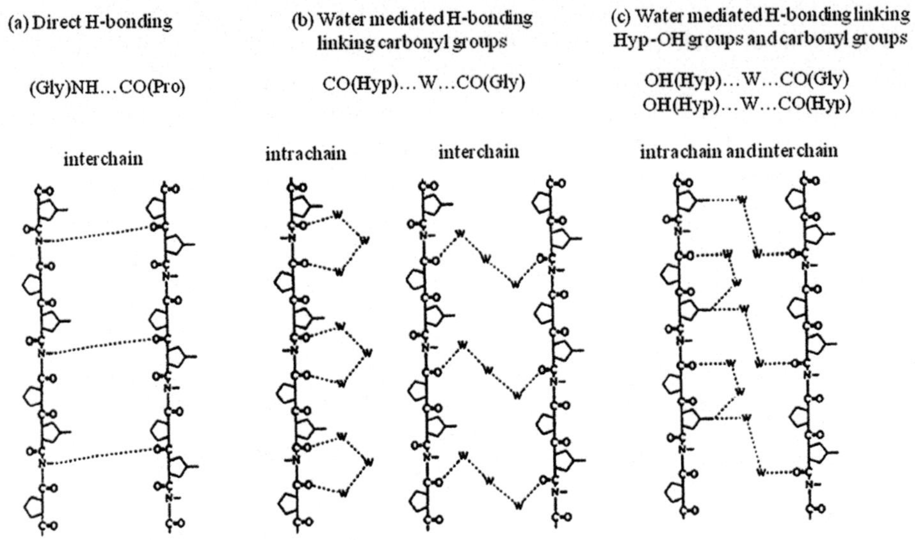

Fig. 5.4 Schematic illustration of the hydrogen bonding patterns stabilizing the collagen structure: **a** direct hydrogen bonding; **b** water mediated hydrogen bonding linking carbonyl groups and **c** water mediated hydrogen bonding linking hydroxyproline OH and carbonyl groups. W: water. *Adapted from* [20] *with permission from John Wiley and Sons, Inc*

5.2 Biopolymers in Drug Delivery

Biopolymers have been widely used, as excipients, in traditional drug formulation - tablets and capsules. Excipients are inactive ingredients which are mixed with active pharmaceutical ingredients in order to create a drug product that is ready for a specified application. They serve various purposes, such as tablet binder, filling agent, coating agent, and emulsifying agent [21]. Hard capsules are usually gelatin, cellulose, and pullulan based. Some of the biopolymers also exhibit pharmacological activities. For example, polyanionic compounds, such as sulphated polysaccharides, may inhibit certain viruses by limiting the absorption of viruses onto the surface of host cells [21].

Biopolymers, especially cationic biopolymers, have been widely used for gene delivery. Some of the commonly used cationic biopolymers for gene delivery include poly(amino acids) and chitosan [22–24]. The mechanism of cationic polymer-assisted gene delivery is shown in Fig. 5.5. The first stage is generally incubating genetic material with cationic polymer-nucleic acid polyplexes (due to ionic interactions). The most accepted theory for releasing nucleic acids from the endosome/lysosome is the proton sponge effect. The cationic groups, usually amine, sequester protons, acidifying the endosome/lysosome. The "proton pump" process leads to the retention of water molecules resulting in endosome/lysosome swelling and rupture [25]. Take chitosan and its derivatives for example, they have been widely studied for gene delivery. Chitosan derivatives, in particular, offer improved solubility and other advantages for gene delivery [26]. One of such derivatives is amphiphilic chitosan modified by myristic acid and tertiary amino groups (Fig. 5.6). The results showed a significant increase in gene complexing capacity and decreased carrier amount for achieving the highest transfection efficiency [27].

Biopolymers have been used to design different types of drug delivery systems—prodrugs, diffusion controlled, erodible, and osmotic controlled drug delivery systems (Fig. 5.7). Moreover, novel drug delivery systems are generally engineered with targetability and/or stimuli-responsiveness. Targeting can be achieved through specific biopolymer—cell interactions. For biopolymers, they can be used to design pH, temperature,

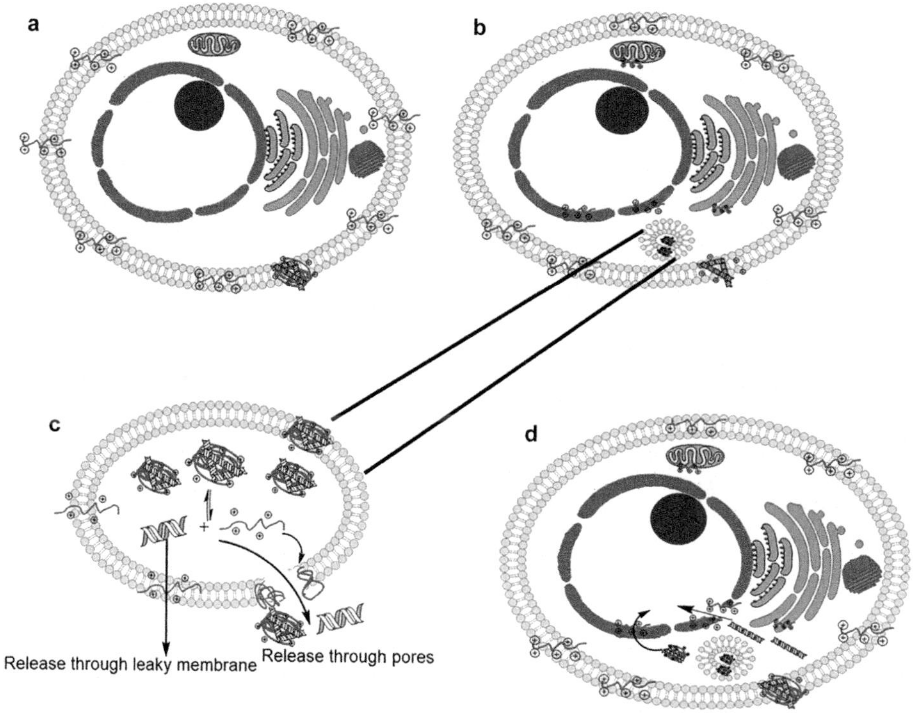

Fig. 5.5 Mechanism of cationic polymer-assisted gene delivery. **a** Free cationic polymer intercalates into the plasma membrane and/or is released from associated polyplexes. **b** Cationic polymer is dispersed into internal cellular membranes via lipid recycling pathways. Polyplexes are endocytosed. **c** Cationic polymer intercalated into endosomal membranes induces membrane permeability and facilitates release of genetic material into the cytosol. **d** Cationic polymer associated with the nuclear membrane via lipid recycling and/or cytosolic pathways induces permeabilization and facilitates transport into the nucleus (hypothesized). W: water. *Adapted with permission from* [24]. *Copyright (2016) American Chemical Society*

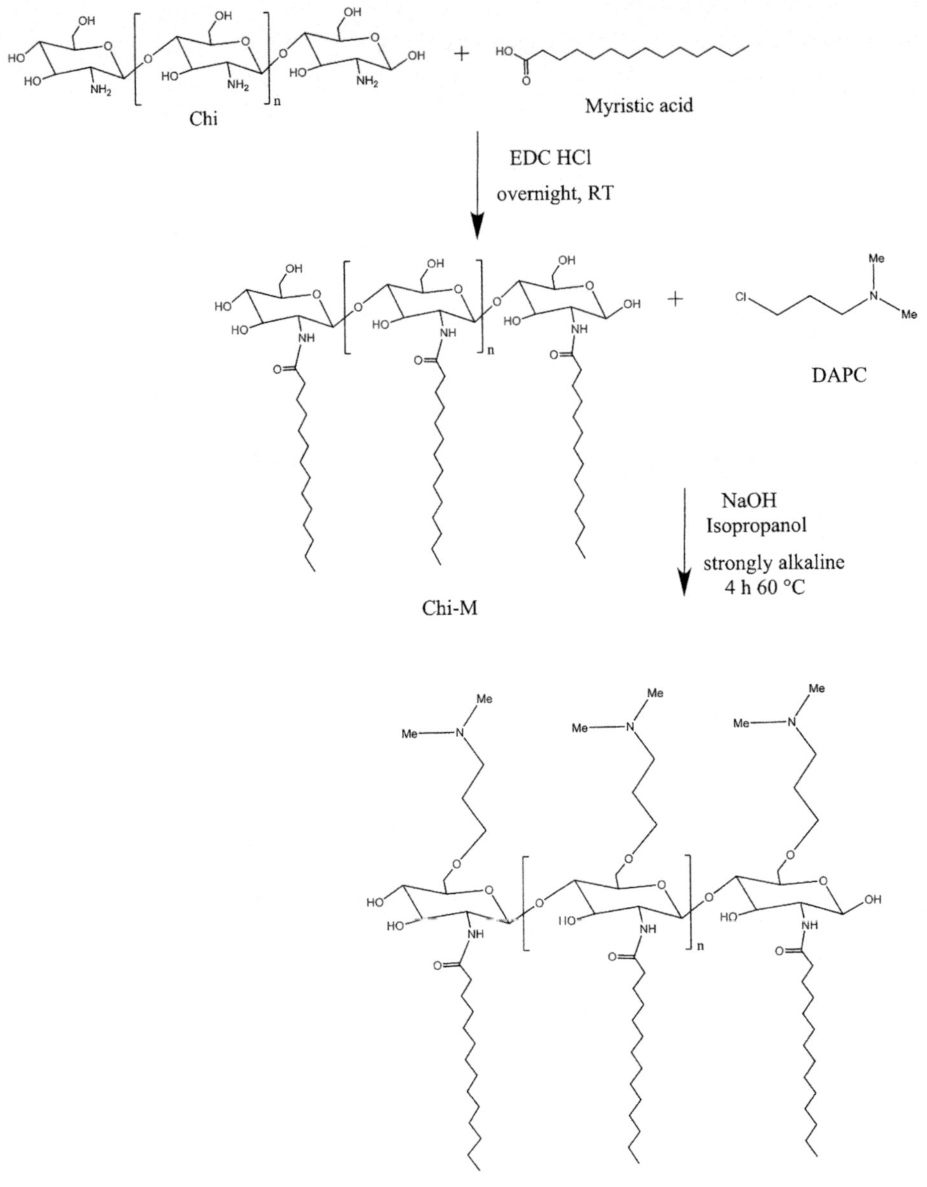

Fig. 5.6 Reaction scheme of the molecular structure formed by myristic acid and DAPC modification of the chitosan (Chi) chain. EDC HCl: 1-Ethyl-3-(3-dimethylaminopropyl) carbodiimide hydrochloride, DAPC: 3-dimethylamino-1-propyl chloride hydrochloride. *Adapted form* [27] *with permission from Elsevier*

photo-sensitive and other stimuli-responsive delivery systems. For novel biopolymer-based drug delivery systems they are usually at micro and nano-scale and in the form of particles (spheres), capsules, micelles, and more (Fig. 5.8) [28, 29]. Novel drug delivery systems are also utilized for controlled drug delivery, such as delayed (a dosage form that releases a discrete portion or portions of drug at a time other than promptly after administration) and extended (a dosage form that allows at least a twofold reduction in dosage frequency as compared to that drug presented as an immediate-release (conventional) dosage form; such as sustained-release (drug release at a programmed rate)) drug delivery (Fig. 5.8) [30].

Regarding micro-scale polymeric systems, microspheres/capsules have been commonly used for different routes of administration. For example, pectin-based microspheres have been designed for oral and pulmonary drug delivery. Calcium-Oligochitosan-Pectin microspheres are responsive to both pH and colonic enzymes and show great potential for colonic drug delivery (Fig. 5.9) [31]. Moreover, gelatin-coated pectin microspheres (~ 3 μm in diameter) were prepared for pulmonary drug delivery using an electrospray setup. The microspheres showed high responsiveness to the stimulated lung fluid (Fig. 5.10) [32]. Nanoscale carriers have been the major focus for developing novel drug delivery systems which significantly contribute to the rise of the nanomedicine. A large percent of these systems is designed for cancer treatment. These carriers are often multi-functional—such

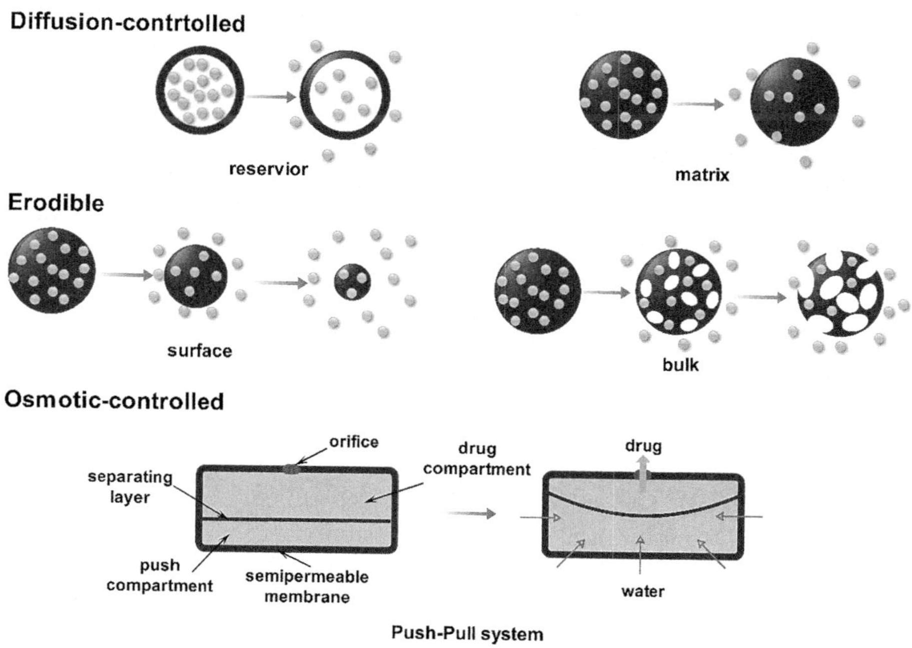

Fig. 5.7 Different controlled drug delivery systems

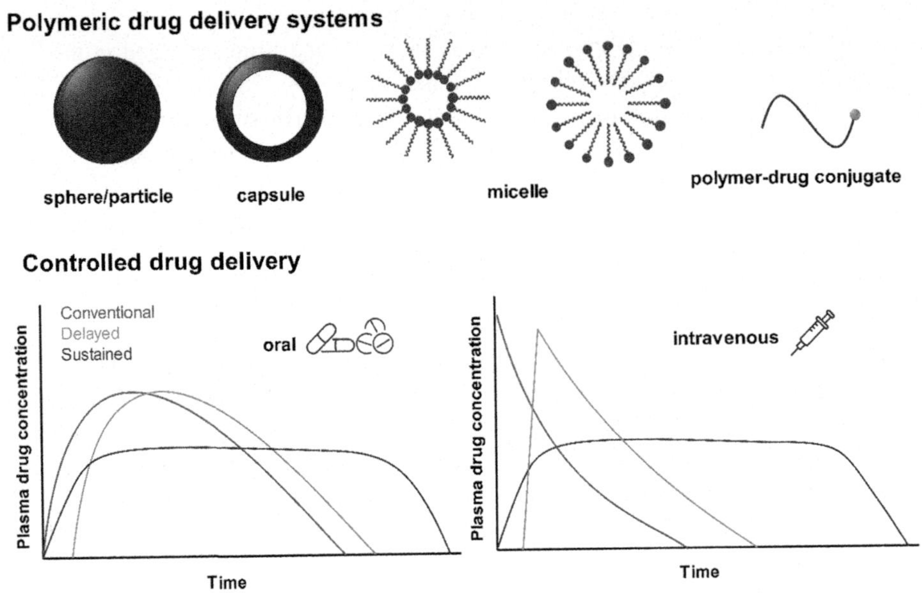

Fig. 5.8 Different polymeric drug delivery systems (top) and controlled drug delivery (bottom)

as targeted drug delivery with bioimaging capacity. Nanoscale carriers can also be used for passive targeting as the endothelial cells in the tumor vasculature can range from 200–2000 nm (depending on the tumor type, location, and environment) as well as the poor lymphatic function. Hence, the nanoscale carriers are not rapidly cleared but accumulate in the tumor interstitium (i.e., enhanced permeability and retention (EPR)) [33]. Carriers can also be designed to be responsive to internal and/or external stimuli improving the accumulation at the desired site (e.g., apply an external magnetic field) and enhancing the drug release (releasing drug in response to pH, temperature, light, etc.) (Fig. 5.11) [33, 34]. Moreover, PEGylation (the attachment of poly(ethylene glycol)) is commonly used to improve the performance of nanocarrier (such as increase the circulation time); however, it may lead to the accelerated blood clearance (i.e., ABC phenomena) upon repeated administration involving anti-PEG antibodies production at the first injection [35, 36].

5.3 Biopolymers in Cellular Therapy and Tissue Engineering

Regarding cellular therapy applications, biopolymers often provide support and protection for cells, such as through cell encapsulation. Cell encapsulation (usually at macro or micro-scale) is the process of entrapping cells (usually allogenic or xenogeneic) into a matrix, mostly spherical in shape and in the form of a polymeric hydrogel [37, 38].

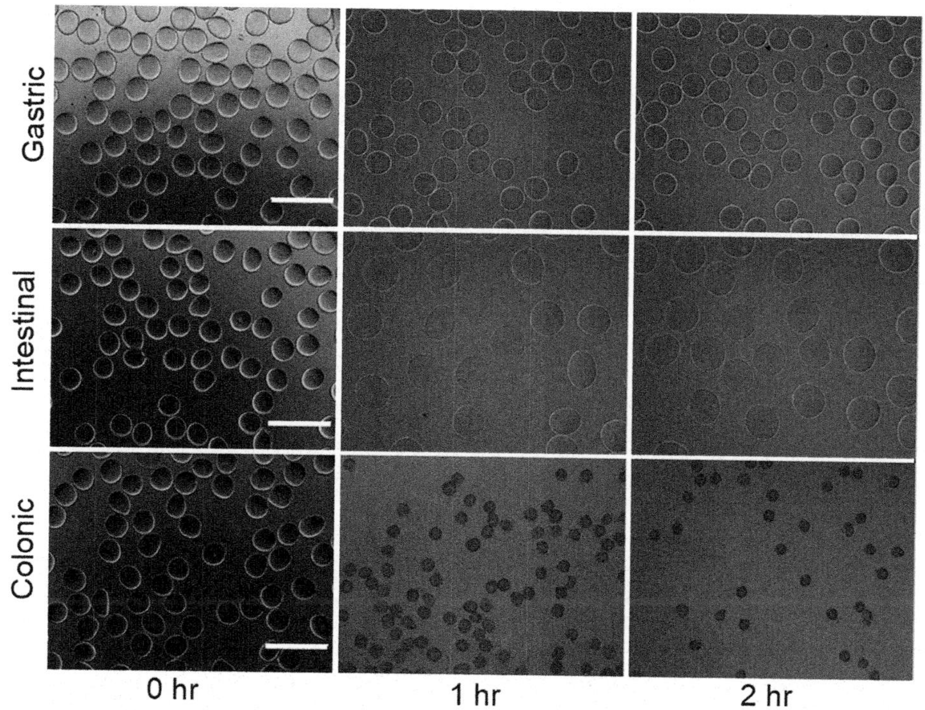

Fig. 5.9 Responsiveness of calcium-oligochitosan-pectin microspheres (incubated in cross-linking solutions for 15 min) to simulated gastrointestinal environments. Scale bar represents 1000 μm. *Adapted from*[31] *with permission from Taylor and Francis*

The polymeric matrix/membrane serves as an immune-isolating barrier while controlling the release of therapeutic agents secreted by encapsulated cells (Fig. 5.12) [39, 40]. Cell encapsulation was first used for diabetes treatment and then for neurological, musculoskeletal, and cardiovascular diseases, as well as liver failure [40–43]. Alginate is the most-used biopolymer for cell encapsulation. There are many alginate-based cell encapsulation systems that have been developed including alginate-chitosan-alginate microcapsules, alginate-PLL-alginate microcapsules, alginate-PLO-alginate microcapsules, Ba-alginate microbeads, and more [42, 44]. The major challenge currently for the clinical application of encapsulated cells for cellular therapy is the immune response generated against the polymeric matrix/membrane and their contents [38]. Besides for cellular therapy, cell encapsulation has also been used for cell differentiation [45–47] and cell banking [40, 48, 49].

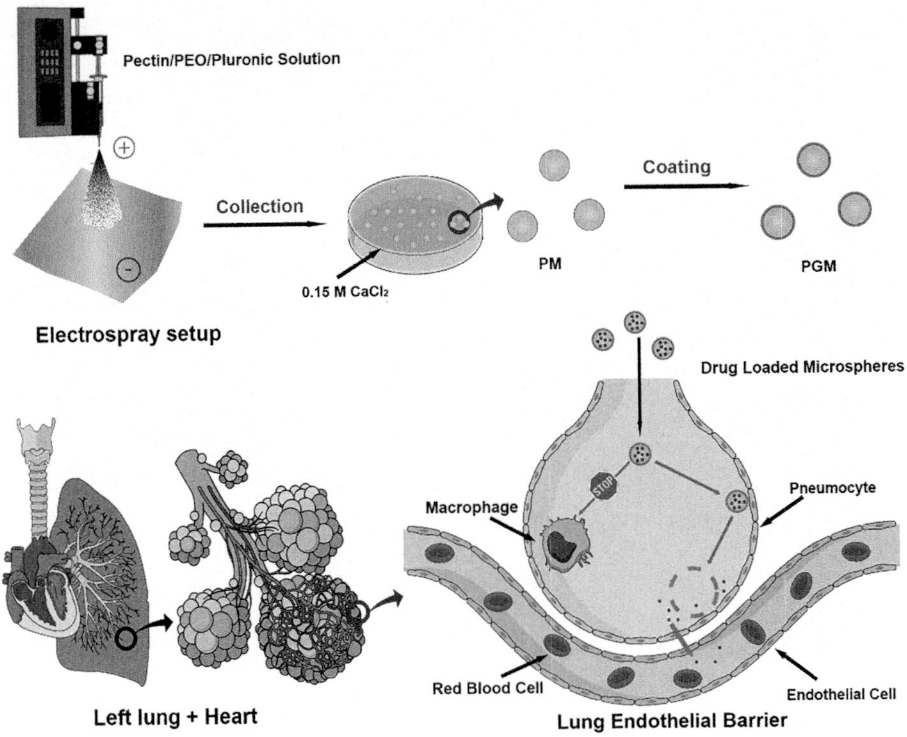

Fig. 5.10 Schematic of gelatin-coated pectin microspheres (PGM) preparation process (top) and mechanism of pulmonary drug delivery/release (bottom). PM: Ca-Pectin microspheres. *Adapted from* [32]

According to U.S. Health Resources and Services Administration, 17 people die each day waiting for an organ transplant [50]. To seal the gap between the numbers of transplant waitlist and organ donation (Fig. 5.13), tissue engineering offers an alternative approach. Collagen has been widely to generate matrix for cells towards tissue engineering, considering it is the major component of actual extracellular matrix [51]. There are various collagen-based scaffold products in market, such as SpongeCol® columnar pore architecture discs, CollaFibR™, Koken's atelocollagen honeycomb sponge, and Avitene™ Ultrafoam™ Collagen Sponge.

Bioprinting is the application of additive manufacturing that involves building a tissue or organ layer-by-layer using bottoms-up approach (recapitulating the physiologically relevant architecture of tissue/organ). One of the advantages of bioprinting is its potential for personalized medicine [52–54]. Although there are different bioprinting approaches,

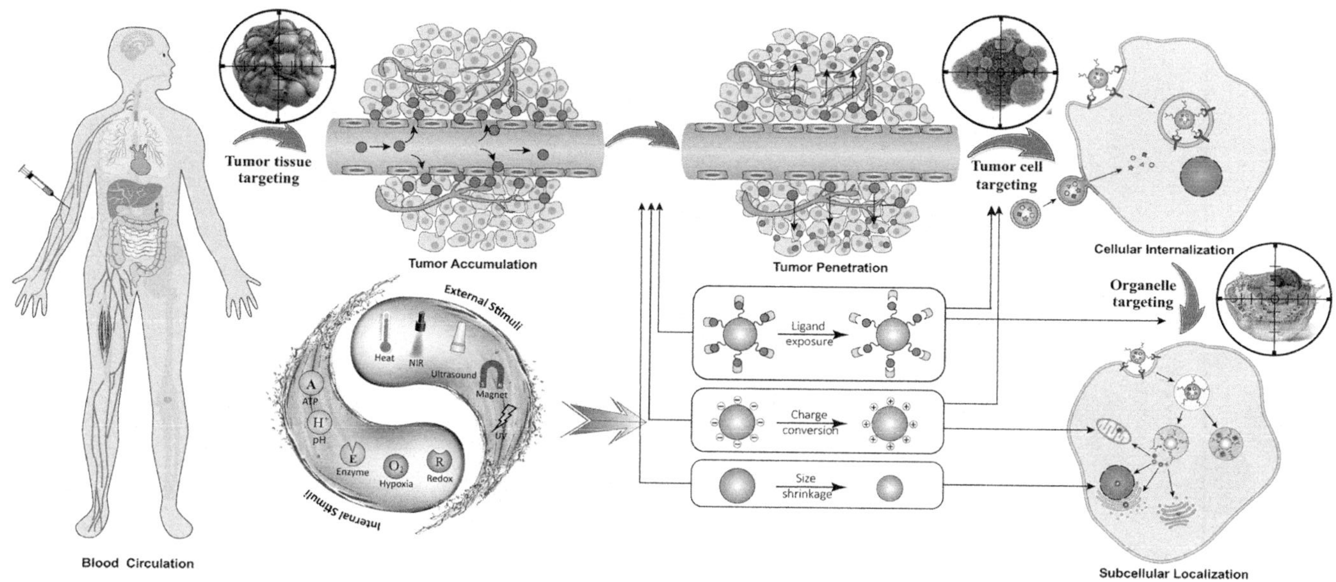

Fig. 5.11 Maximizing therapeutic outcomes of cancer nanomedicine through the stimuli-triggered dynamic integration of multistage tumor targeting. Either endogenous (such as changes in pH) or exogenous (such as variations in temperature) stimuli can be utilized to trigger nanocarrier change (such as size reduction) and therefore can dynamically integrate tumor tissue-, cell- and organelle-specific targeting to sequentially achieve high tumor accumulation, deep tumor penetration, efficient cellular internalization, and accurate organelle localization. *Adapted from* [34]

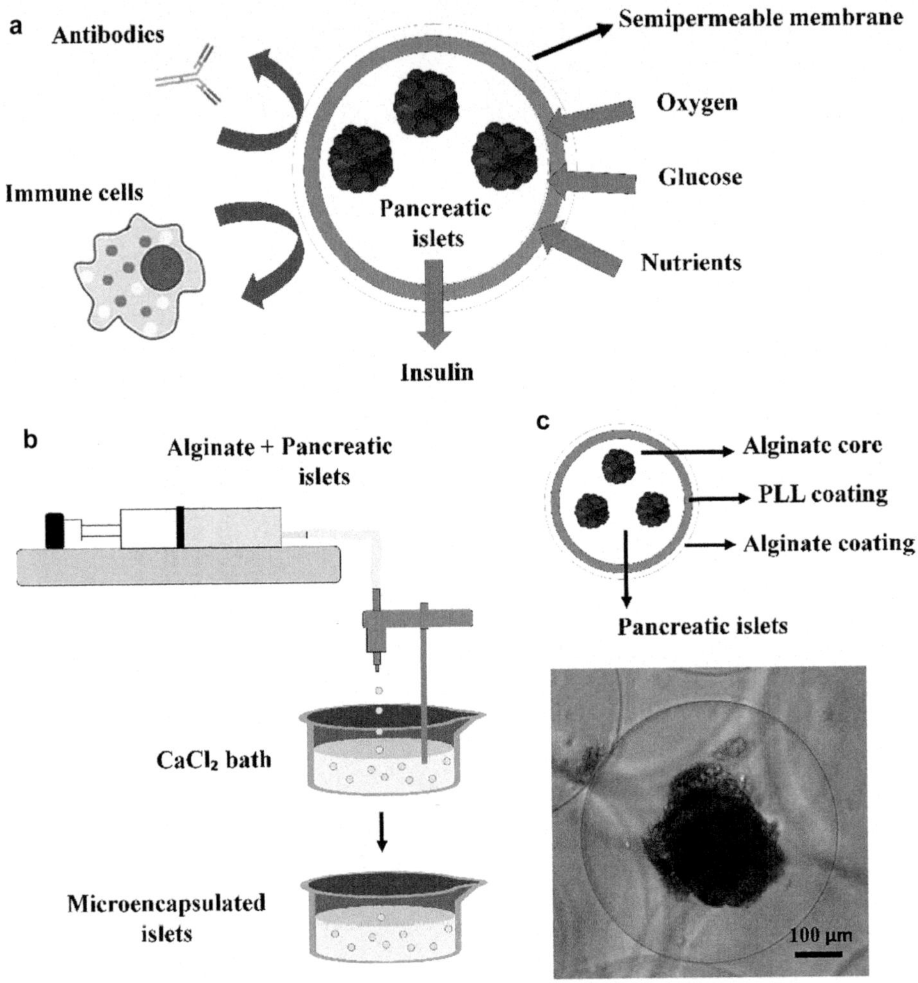

Fig. 5.12 **a** Principle of immunoisolation by a microcapsule. **b** Schematic of pancreatic islet microencapsulation within alginate microcapsules obtained by ionic cross-linking in a $CaCl_2$ bath where droplets that are generated by an electrostatic droplet generator are gelled. **c** Layers structure of an alginate-poly-L-lysine-alginate microcapsule and, brightfield image of a microencapsulated pseudoislet generated from the rat insulinoma INS1E cell line. PLL: poly-L-lysine. Scale bar: 100 µm. *Adapted from* [39]

such as microextrusion, stereolithography, laser-assisted, inkjet-based, and microfluidic bioprinting, the process for generating artificial tissue/organ using bioprinting is typical. As shown in Fig. 5.14, imaging is the first step (for personalized medicine in particular) to generate the structure design. Different design approaches, such as biomimicry, self-assembly, and mini-tissues, can be utilized (singly or in combination) [55]. Cells and

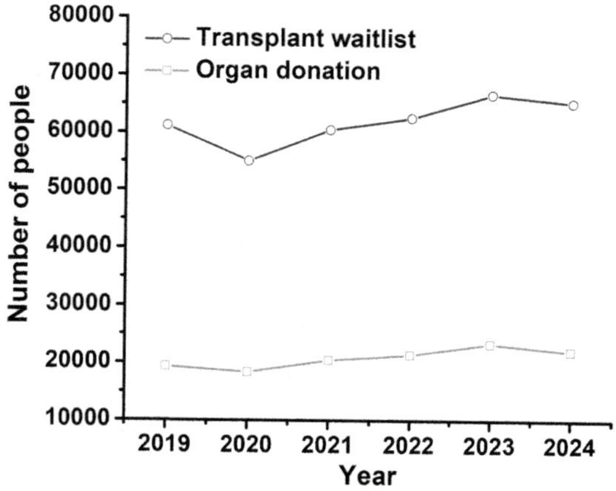

Fig. 5.13 Organ donation and transplant waitlist statistics. Data from: https://data.hrsa.gov/topics/health-systems/organ-donation

biomaterials (i.e., bioink) are needed for bioprinting. Polymeric materials and extracellular matrix (such as from decellularized tissue) are used for bioprinting materials. There are many types of commercial biopolymer-based bioink, such as TissueFab® bioinks and CELLINK Vivoink™ medical grade bioink. Once printed, it is critical to let the tissue/organ mature (bioreactors especially those allow for applying specific stimuli are necessary) [56]. Interestingly, biopolymer-based microspheres have been shown to promote vascularization for vascularized tissue engineering [57–59]. A pectin-based bioink incorporating microspheres have been developed [60–62]. The bioink base was pectin and Pluronic® F127. As shown in Fig. 5.15, the printing bed was heated to 37 °C and the first few layers were generated by the solidification of Pluronic® F127 (due to its unique thermal responsiveness). Then sequential layers were gelled by introducing the warm $CaCl_2$ solution (formation of Ca-pectin hydrogel) [63, 64]. When it comes to the tissue engineering applications, the base bioink needs be mixed with cells (both functional (such as stem cells) and endothelial cells) and microspheres. Microspheres can also be loaded with drug molecules [60, 61, 65]. The design has been used as an injectable scaffold for endometrial regeneration with microspheres also served as the controlled drug delivery/release carrier [65] (Fig. 5.16).

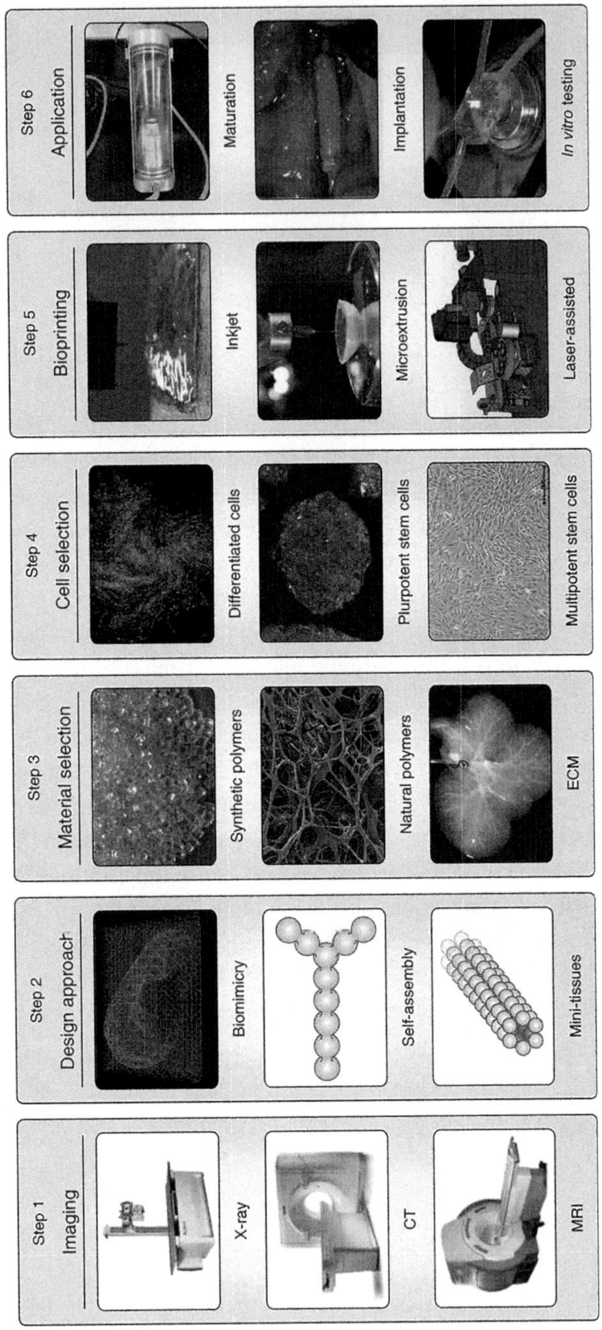

Fig. 5.14 A typical process for bioprinting 3D tissues. *Adapted from* [55] *with permission from Springer Nature*

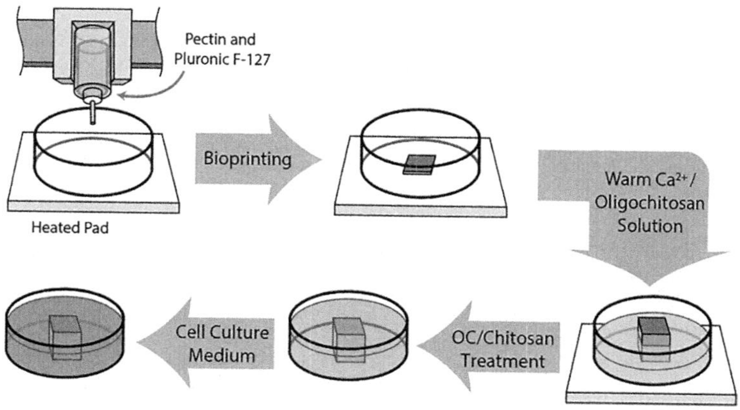

Fig. 5.15 A schematic illustration of the bioprinting process and subsequent post-printing modification. *Copyright © 2019 by Stealey, S. et.al. Adapted from* [63] *by permission of Sage Publications*

Fig. 5.16 Applications of microsphere-incorporated bioink for fabrication of vascularized tissue. BMSCs: bone marrow mesenchymal stem cells, HUVECs: human umbilical vein endothelial cells. *Adapted from* [61]

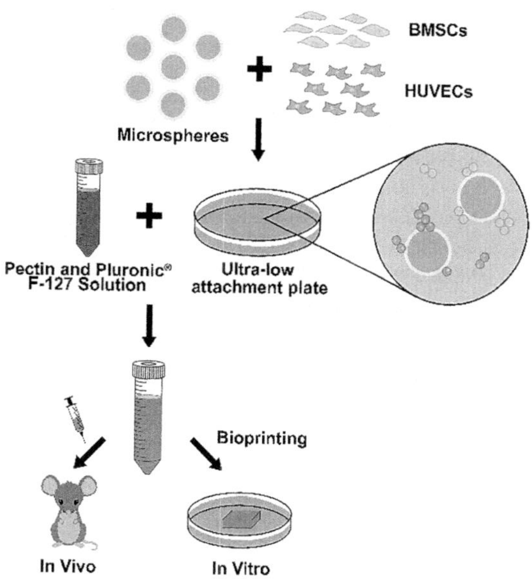

References

1. Baranwal J, Barse B, Fais A, et al. Biopolymer: A Sustainable Material for Food and Medical Applications. Polymers. 2022;14(5).
2. Chopra H. Biopolymers for Wound Healing Applications. Biopolymers for Biomedical Applications. Scrivener Publishing; 2024. p. 393–415.

3. Chopra H, Sharma S, Shabbir M, et al. Future Perspectives of Biopolymers for Biomedical Applications. Biopolymers for Biomedical Applications. Scrivener Publishing; 2024. p. 509–530.
4. Kappen J, John FA, Jose AJ. Biopolymers in Biosensors. Biopolymers for Biomedical Applications. Scrivener Publishing; 2024. p. 439–457.
5. Singh R, Ahn Y-H. Progress of Biopolymers in 3D and 4D Printing for Biomedical Applications. Biopolymers for Biomedical Applications. Scrivener Publishing; 2024. p. 479–507.
6. Soman R, Mangla D, Annu. Bioplastics in Medical Devices. Biopolymers for Biomedical Applications. Scrivener Publishing; 2024. p. 417–437.
7. Biomaterials. An Introduction To Tissue-Biomaterial Interactions. Wiley-Liss; 2002. p. 1–13.
8. Visan AI, Negut I. Development and Applications of PLGA Hydrogels for Sustained Delivery of Therapeutic Agents. Gels. 2024;10(8):497.
9. Samir A, Ashour FH, Hakim AAA, et al. Recent advances in biodegradable polymers for sustainable applications. npj Materials Degradation. 2022;6(1):68.
10. Kreua-ongarjnukool N, Soomherun N, Niyomthai ST, et al. Aliphatic Polyester Nanoparticles for Drug Delivery Systems. In: Ahmad U, Haider MF, Akhtar J, editors. Smart Drug Delivery: IntechOpen; 2021.
11. Hua Y, Su Y, Zhang H, et al. Poly(lactic-co-glycolic acid) microsphere production based on quality by design: a review. Drug Delivery. 2021;28(1):1342–1355.
12. Su Y, Zhang B, Sun R, et al. PLGA-based biodegradable microspheres in drug delivery: recent advances in research and application. Drug Delivery. 2021;28(1):1397–1418.
13. Harugade A, Sherje AP, Pethe A. Chitosan: A review on properties, biological activities and recent progress in biomedical applications. Reactive and Functional Polymers. 2023;191:105634.
14. Tricol Biomedical. OneStop® Bandage & Bandage PRO. Available from: https://tricolbiomedical.com/wp-content/uploads/2023/09/MMF-289_Rev_1.pdf
15. Tang C, Zhou K, Zhu Y, et al. Collagen and its derivatives: From structure and properties to their applications in food industry. Food Hydrocolloids. 2022;131:107748.
16. Ahmad MI, Li Y, Pan J, et al. Collagen and gelatin: Structure, properties, and applications in food industry. International Journal of Biological Macromolecules. 2024;254:128037.
17. Zhang X, Xu S, Shen L, et al. Factors affecting thermal stability of collagen from the aspects of extraction, processing and modification. Journal of Leather Science and Engineering. 2020;2(1):19.
18. Blanco A, Blanco G. Chapter 3 - Proteins. In: Blanco A, Blanco G, editors. Medical Biochemistry: Academic Press; 2017. p. 21–71.
19. Brodsky B, Persikov AV. Molecular Structure of the Collagen Triple Helix. Advances in Protein Chemistry. Vol. 70: Academic Press; 2005. p. 301–339.
20. Nguyen TT, Happillon T, Feru J, et al. Raman comparison of skin dermis of different ages: focus on spectral markers of collagen hydration. Journal of Raman Spectroscopy. 2013;44(9):1230–1237.
21. Karolewicz B. A review of polymers as multifunctional excipients in drug dosage form technology. Saudi Pharmaceutical Journal. 2016;24(5):525–536.
22. Yang W, Mixich L, Boonstra E, et al. Polymer-Based mRNA Delivery Strategies for Advanced Therapies. Advanced Healthcare Materials. 2023;12(15):2202688.
23. Egorova KS, Posvyatenko AV, Larin SS, et al. Ionic liquids: prospects for nucleic acid handling and delivery. Nucleic Acids Research. 2021;49(3):1201–1234.
24. Vaidyanathan S, Orr BG, Banaszak Holl MM. Role of Cell Membrane–Vector Interactions in Successful Gene Delivery. Accounts of Chemical Research. 2016;49(8):1486–1493.

25. Nel AE, Mädler L, Velegol D, et al. Understanding biophysicochemical interactions at the nano–bio interface. Nature Materials. 2009;8(7):543–557.
26. Dong L, Li Y, Cong H, et al. A review of chitosan in gene therapy: Developments and challenges. Carbohydrate Polymers. 2024;324:121562.
27. Eker Fidan EB, Bal K, Şentürk S, et al. Enhancing gene delivery efficiency with amphiphilic chitosan modified by myristic acid and tertiary amino groups. International Journal of Biological Macromolecules. 2024;282:136775.
28. Li W, Tang J, Lee D, et al. Clinical translation of long-acting drug delivery formulations. Nature Reviews Materials. 2022;7(5):406–420.
29. Syed MH, Zahari MAKM, Khan MMR, et al. An overview on recent biomedical applications of biopolymers: Their role in drug delivery systems and comparison of major systems. Journal of Drug Delivery Science and Technology. 2023;80:104121.
30. Shargel L, Wu-Pong S, Yu ABC. Chapter 17. Modified-Release Drug Products. Applied Biopharmaceutics & Pharmacokinetics, 6e. New York, NY: The McGraw-Hill Companies; 2012.
31. Stealey S, Guo X, Majewski R, et al. Calcium-oligochitosan-pectin microcarrier for colonic drug delivery. Pharmaceutical Development and Technology. 2020;25(2):260–265.
32. Chai A, Schmidt K, Brewster G, et al. Design of Pectin-Based Hydrogel Microspheres for Targeted Pulmonary Delivery. Gels. 2023;9(9):707.
33. Navya PN, Kaphle A, Srinivas SP, et al. Current trends and challenges in cancer management and therapy using designer nanomaterials. Nano Convergence. 2019;6(1):23.
34. Fan D, Cao Y, Cao M, et al. Nanomedicine in cancer therapy. Signal Transduction and Targeted Therapy. 2023;8(1):293.
35. Gao Y, Joshi M, Zhao Z, et al. PEGylated therapeutics in the clinic. Bioeng Transl Med. 2024 Jan;9(1):e10600.
36. Tenchov R, Sasso JM, Zhou QA. PEGylated Lipid Nanoparticle Formulations: Immunological Safety and Efficiency Perspective. Bioconjugate Chemistry. 2023;34(6):941–960.
37. Zhang W. Encapsulation of Transgenic Cells for Gene Therapy. In: Hashad D, editor. Gene Therapy—Principles and Challenges: IntechOpen; 2015.
38. Ashimova A, Yegorov S, Negmetzhanov B, et al. Cell Encapsulation Within Alginate Microcapsules: Immunological Challenges and Outlook. Frontiers in Bioengineering and Biotechnology. 2019;7.
39. Espona-Noguera A, Ciriza J, Cañibano-Hernández A, et al. Review of Advanced Hydrogel-Based Cell Encapsulation Systems for Insulin Delivery in Type 1 Diabetes Mellitus. Pharmaceutics. 2019;11(11):597.
40. Zhang W, He X. Microencapsulating and Banking Living Cells for Cell-Based Medicine. Journal of Healthcare Engineering. 2011;2(4):875406.
41. Yuan X, Wu J, Sun Z, et al. Preclinical efficacy and safety of encapsulated proliferating human hepatocyte organoids in treating liver failure. Cell Stem Cell. 2024;31(4):484–498.e5.
42. Orive G, Santos E, Poncelet D, et al. Cell encapsulation: technical and clinical advances. Trends in Pharmacological Sciences. 2015;36(8):537–546.
43. Basta G, Montanucci P, Calafiore R. Microencapsulation of cells and molecular therapy of type 1 diabetes mellitus: The actual state and future perspectives between promise and progress. J Diabetes Investig. 2021 Mar;12(3):301–309.
44. Zhang W, Zhao S, Rao W, et al. A novel core–shell microcapsule for encapsulation and 3D culture of embryonic stem cells. Journal of Materials Chemistry B. 2013;1(7):1002–1009.
45. Agarwal P, Zhao S, Bielecki P, et al. One-step microfluidic generation of pre-hatching embryo-like core–shell microcapsules for miniaturized 3D culture of pluripotent stem cells. Lab on a Chip. 2013;13(23):4525–4533.

46. Zhao S, Agarwal P, Rao W, et al. Coaxial electrospray of liquid core–hydrogel shell microcapsules for encapsulation and miniaturized 3D culture of pluripotent stem cells. Integrative Biology. 2014;6(9):874–884.
47. Zhang W, Zhao S, He X. Proliferation and differentiation of mesenchymal stem cells encapsulated in miniaturized 3D liquid core of alginate-chitosan-alginate (ACA) microcapsules. Archives of Stem Cell Research. 2015;2(1):1004.
48. Zhang W, Yang G, Zhang A, et al. Preferential vitrification of water in small alginate microcapsules significantly augments cell cryopreservation by vitrification. Biomedical Microdevices. 2010;12(1):89–96.
49. Huang H, He X, Yarmush ML. Advanced technologies for the preservation of mammalian biospecimens. Nature Biomedical Engineering. 2021;5(8):793–804.
50. U.S. Health Resources & Services Administration. Organ Donation Statistics. Available from: https://www.organdonor.gov/learn/organ-donation-statistics
51. Zhou H, Li W, Pan L, et al. Human extracellular matrix (ECM)-like collagen and its bioactivity. Regenerative Biomaterials. 2024;11.
52. Agarwal S, Saha S, Balla VK, et al. Current Developments in 3D Bioprinting for Tissue and Organ Regeneration–A Review. Frontiers in Mechanical Engineering. 2020;6.
53. Lam EHY, Yu F, Zhu S, et al. 3D Bioprinting for Next-Generation Personalized Medicine. Int J Mol Sci. 2023 Mar 28;24(7).
54. Jo Y, Hwang DG, Kim M, et al. Bioprinting-assisted tissue assembly to generate organ substitutes at scale. Trends in Biotechnology. 2023;41(1):93–105.
55. Murphy SV, Atala A. 3D bioprinting of tissues and organs. Nature Biotechnology. 2014;32(8):773–785.
56. Saini G, Segaran N, Mayer JL, et al. Applications of 3D Bioprinting in Tissue Engineering and Regenerative Medicine. Journal of Clinical Medicine. 2021;10(21):4966.
57. Zhang W, Choi JK, He X. Engineering Microvascularized 3D Tissue Using Alginate-Chitosan Microcapsules. J Biomater Tissue Eng. 2017;7(2):170–173.
58. Yao R, Alkhawtani AYF, Chen R, et al. Rapid and efficient in vivo angiogenesis directed by electro-assisted bioprinting of alginate/collagen microspheres with human umbilical vein endothelial cell coating layer. Int J Bioprint. 2019;5(2.1):194.
59. Wang W, Liu Y, Liu Z, et al. Evaluation of Interleukin-4-Loaded Sodium Alginate–Chitosan Microspheres for Their Support of Microvascularization in Engineered Tissues. ACS Biomaterials Science & Engineering. 2021;7(10):4946–4958.
60. Johnson DL, Ziemba RM, Shebesta JH, et al. Design of pectin-based bioink containing bioactive agent-loaded microspheres for bioprinting. Biomedical Physics & Engineering Express. 2019;5(6):067004.
61. Kanungo M, Wang Y, Hutchinson N, et al. Development of Gelatin-Coated Microspheres for Novel Bioink Design. Polymers. 2021;13(19):3339.
62. Zieman J, Cohan M, Wang Y, et al. Development of Gelatin-Coated Hydrogel Microspheres for Novel Bioink Design: A Crosslinker Study. Pharmaceutics. 2023;15(1):90.
63. Stealey S, Guo X, Ren L, et al. Stability improvement and characterization of bioprinted pectin-based scaffold. Journal of Applied Biomaterials & Functional Materials. 2019;17(1):2280800018807108.
64. Banks A, Guo X, Chen J, et al. Novel bioprinting method using a pectin based bioink. Technology and Health Care. 2017;25:651–655.
65. Wu Y, Gu S, Cobb JM, et al. E2-Loaded Microcapsules and Bone Marrow–Derived Mesenchymal Stem Cells with Injectable Scaffolds for Endometrial Regeneration Application. Tissue Engineering Part A. 2024;30(3-4):115–130.

Biopolymers: Food Industry Applications 6

6.1 Introduction to Food Industry Applications of Biopolymers

Biopolymers play a vital role in the agriculture-food sector (Fig. 6.1). In agriculture, biopolymers have been used in various ways to improve the crop (raw food materials) production mainly. Biopolymers have been applied for seed coatings to prevent them from aging at high temperatures and humidity and promote germination [1]. For example, seeds have been coated with a silk/trehalose inner layer with plant-beneficial microbes and a pectin/carboxymethylcelluose outer layer to increase germination and water-stress tolerance in semi-arid, sandy soils (Fig. 6.2) [2, 3]. Biopolymers have also been used as water retention agents for soil conditioners and carriers of agrochemicals for their controlled release [4]. Moreover, biopolymers serve as alternatives to conventional plastic mulches to reduce pollution and soil degradation [5].

Regarding food applications of biopolymers, they have been utilized in nearly every field of food industry. To start with, major food ingredients from both plant and animal sources, polysaccharides and proteins, are biopolymers. Biopolymers have been used food additives, such as emulsifiers to improve the stability and physicochemical properties of food emulsions [6]. Most recently, the introduction of 3D printing in the food industry has been observed to fulfill the need for personalized nutrition and ensure the sustainability of food supply [7]. There are other benefits of 3D printed foods, such as to help improve the appetites of people with dysphagia [8]. The world's first 3D bioprinted and cultivated ribeye steak was revealed in 2021 using bioprinting and real cow cells [9]. There are also products of plant-based 'alt-steak' and Redefine Meat Ltd (https://www.redefinemeat.com/company/) is one of the pioneers of this field. Soy protein has been used to produce steak-like foods through 3D printing. For example, a recent study showed that textured soybean protein with xanthan gum (reducing the textured soybean protein inks) showed the best printing characteristics and sample integrity after frying. Moreover, it

W. Zhang, *Biopolymer Engineering*, Synthesis Lectures on Chemical Engineering and Biochemical Engineering, https://doi.org/10.1007/978-3-031-98868-4_6

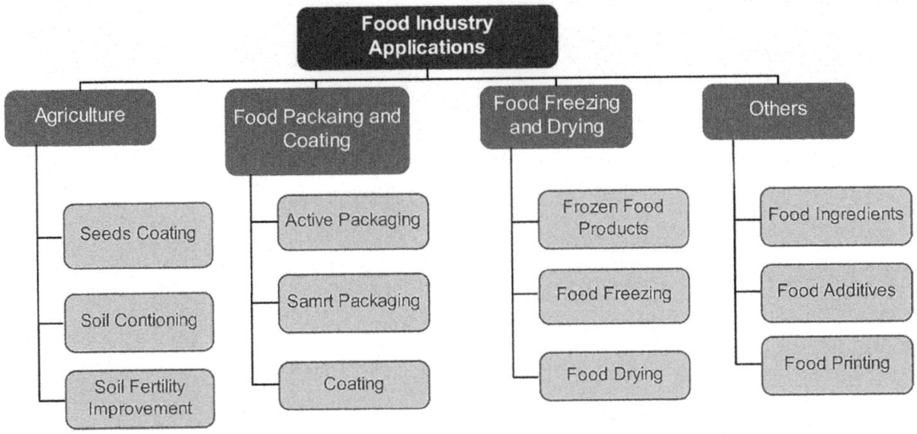

Fig. 6.1 Applications of biopolymers in agriculture-food sector

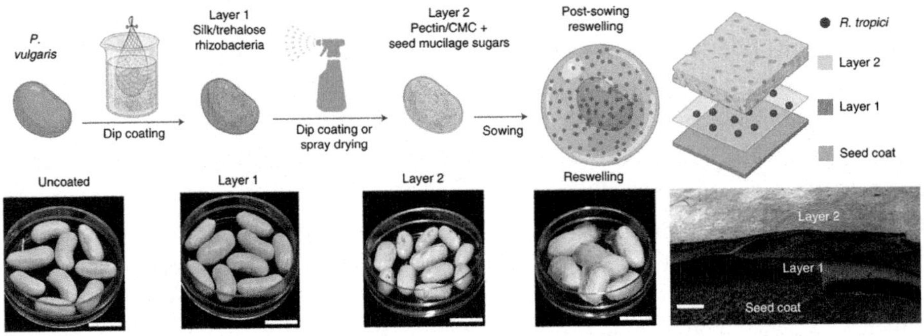

Fig. 6.2 Schematic diagram of the two-layered seed coating fabrication process. Scale bar for the scanning electron microscope image of the coating, 10 μm. *P. vulgaris*: *Phaseolus vulgaris* (common bean), *R. tropicl*: *Rhizobium tropici* CIAT 899 (bacterial). *Adapted from* [3] *with permission from Springer Nature*

was found that different infilling patterns and infill rates had a significant impact on the texture properties (Fig. 6.3) [10]. In this chapter, the focuses are on biopolymers in food packaging, coating, freezing and drying.

6.2 Biopolymers in Food Packaging and Coating

Packaging has an essential role in food quality, safety, and shelf-life extension [10]. Biopolymers have been used for the design of both active and smart (intelligent) packaging compared to traditional food packaging based on paper, glass, metals, aluminum

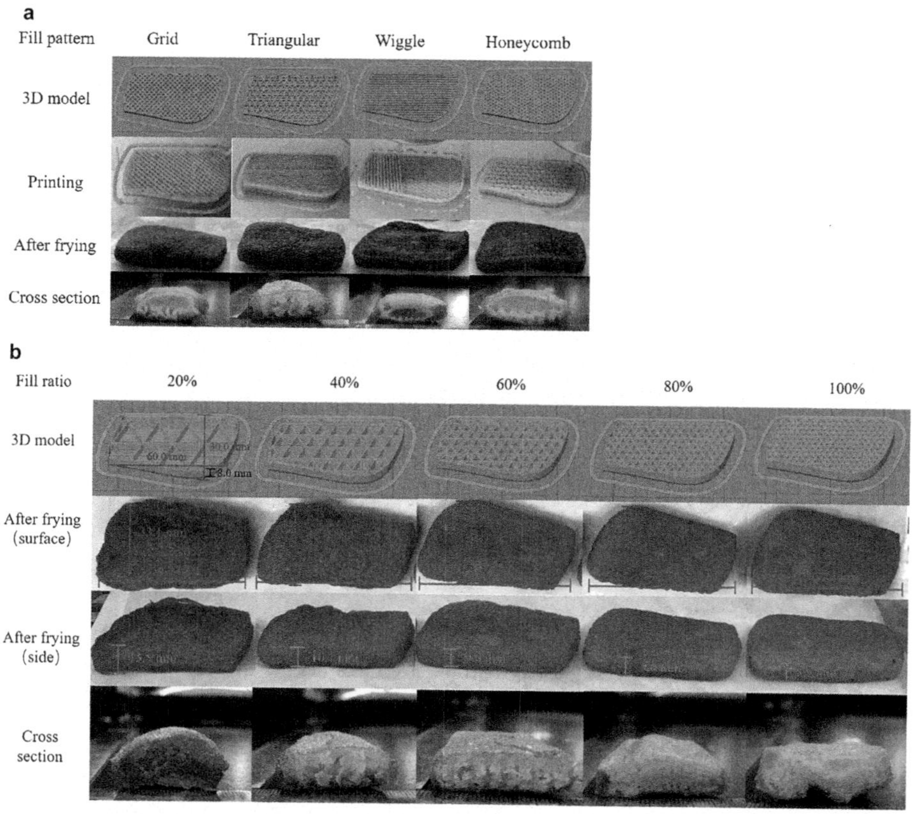

Fig. 6.3 Fried 3D printed steak-like food samples with different infill patterns but same infill ratios (**a**) and with different infill rations (**b**). *Adapted from* [10]

foils, and plastics. Active packaging can be defined as packaging that interacts with food and environment, including packaging able to reduce/inhibit microbial growth, delay oxidation and moisture migration, as well as control the respiration rate of fresh-cut fruit and vegetables [11]. The main goal of active packaging is to prolong the shelf life and the quality of the food [12]. Biopolymers are often used as the base materials for active packaging incorporated with bioactive compounds, such as antioxidants and antimicrobial agents. Commonly used biopolymers in biodegradable active packaging include poly(lactic acid), polyhydroxyalkanoates (PHAs), starch, chitosan, cellulose, rice husks and seaweed [13]. For example, a novel alginate-carboxymethyl cellulose based active packaging for food applications has been developed. The packaging was incorporated with porous vaterite $CaCO_3$ nanoparticles-loaded with cinnamaldehyde (a natural antimicrobial essential oil). And the packaging showed good UV-blocking, water-resistance, and mechanical properties as well as good antibacterial activity [14].

Smart (intelligent) packaging is a type of packaging that acts as "sensors" providing information on packaged food, which can be defined as packaging that is capable of monitoring and communicating information about food quality using time–temperature, ripeness, leakage, physical shock, microbial growth indicators, and so on [11]. For example, chitosan films have been widely studied to develop smart food packaging to monitor food freshness, quality and safety (Fig. 6.4). Sense-active compounds, such as anthocyanins and alizarin, from plant extracts have been utilized to fabricate smart chitosan films; while the common freshness biomarkers considered include ammonia, pH, total volatile basic nitrogen, and thiobarbituric acid reactive substances [15]. One example of such type of film is chitosan/gum Arabic-based emulsion films reinforced with thyme oil encapsulating blood orange anthocyanins. The incorporation of anthocyanin and emulsion enhanced the UV–vis blocking. Thyme oil provided the antioxidant and antibacterial capacity. Anthocyanins change color significantly under different pH environments for visual monitoring the milk spoilage (Fig. 6.5) [16].

Besides packaging, biopolymers have been used for food coating. Coating is commonly used for deep-fried food. Interestingly, the hydrocolloid (composed of gellan gum and guar gum; Fig. 6.6) coating could be used to decrease oil uptake by reducing the heat transfer coefficient [17]. Moreover, coatings, especially edible ones (such as starch-based), have also been used for food preservation as the primary packaging [18]. Biopolymer coverings, such as chitosan, alginate, and gelatin, have emerged as an innovation for extending fruit shelf life [19, 20]. Edible films and coatings have also been investigated for meat and poultry products applications by foaming, dipping, spraying, casting, brushing, individual wrapping and such to increase the shelf-life [12]. For example, Otávio da Mata

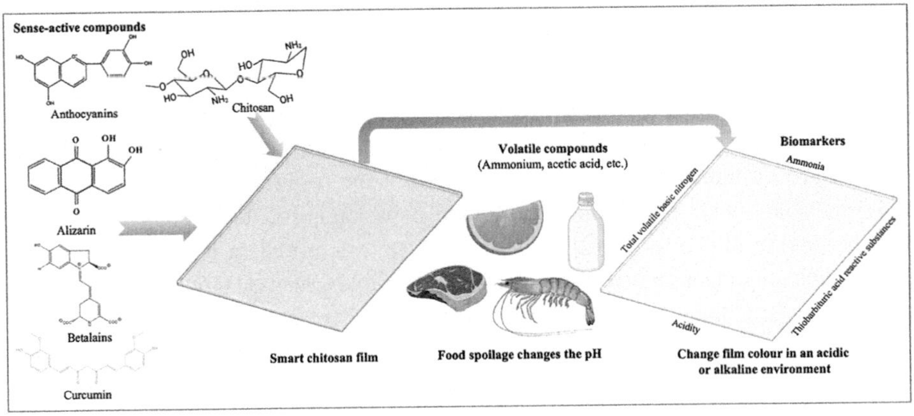

Fig. 6.4 Schematic of developing chitosan film-based smart packaging. Addition of natural sense-active compounds changes the film color. Volatiles compounds are generated during food spoilage. Then, the alteration of the structure of sense-active compounds leading to the film color change. *Adapted from* [15]

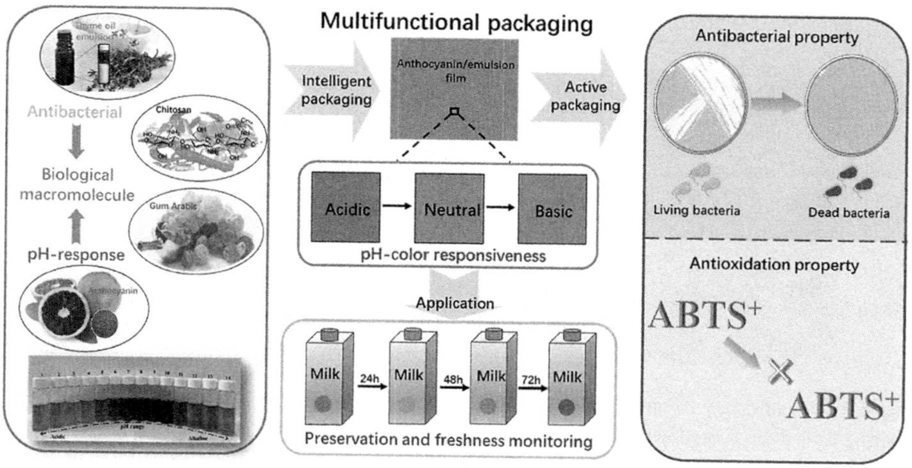

Fig. 6.5 Active chitosan/gum Arabic-based emulsion films reinforced with thyme oil encapsulating blood orange anthocyanins. ABTS: 2,2′-azino-bis-3-ethylbenzthiazoline-6-sulphonic acid. *Adapted from* [16] *with permission from Elsevier*

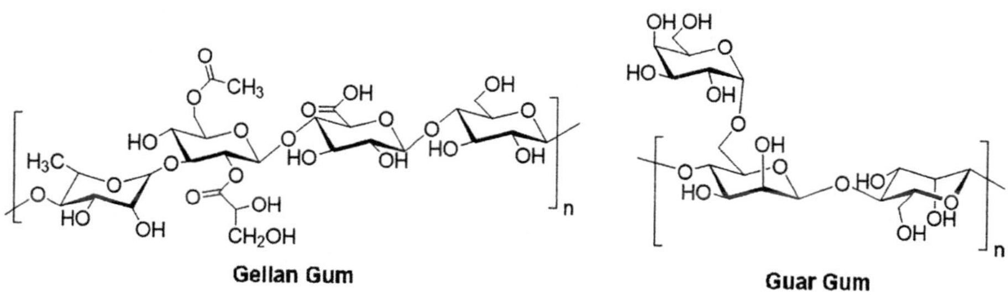

Fig. 6.6 Structure of gellan gum and guar gum

Cunha et al., reported the use of amphiphilic diethylaminoethyl chitosan to produce edible film using casting. The chemical modification of chitosan increased the stretchability and antimicrobial activity of films (Fig. 6.7) [21]. Active compounds, such as antimicrobial compounds, can also be incorporated into the coatings [22].

6.3 Biopolymers in Food Freezing and Drying

Biopolymers have been used as stabilizers, cryoprotectants, and more in food freezing. In ice cream and related products (such as custard), hydrocolloids, such as carboxymethyl cellulose, xanthan gum, and gelatin, serve as stabilizers (also increasing viscosity) to hinder ice crystal growth as temperature fluctuates during storage. Although there is

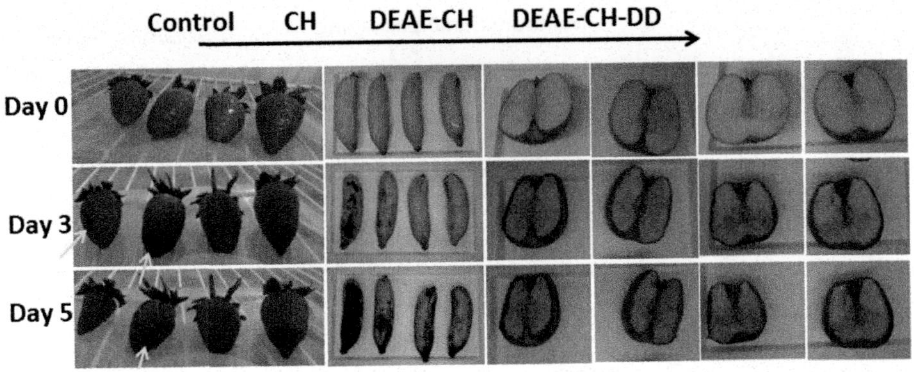

Fig. 6.7 Visual decay of strawberries, bananas and apples during storage time of 5 days without coating and after coated with antimicrobial edible films of CH, DEAE-CH and DEAE-CH-DD (left to right). CH: chitosan, DEAE-CH: diethylaminoethyl chitosan, DEAE-CH-DD: DEAE-CH subjected to an alkylation reaction. Arrows indicate signs of rot/spoilage. *Adapted from* [21] *with permission from Elsevier*

no significant impact on the initial ice crystal size distribution in ice cream, hydrocolloid stabilizers limit the rate of growth of ice crystals during recrystallization [23–25]. Recrystallization is a process in regard to the change of characteristics of ice crystals such as shape, size, number, orientation and perfection [26]. Recrystallization often leads to the increase in crystal size lowering food quality and changing food texture [27]. For ice cream products, ice crystals smaller than 50 μm produce a desirable and creamy texture. The ice cream quality highly depends on the storge condition, as the temperature fluctuation can cause melting and refreezing during storage (main mechanism for recrystallization) [25]. Stabilizers (ι-carrageenan and its acid and enzymatic hydrolyzates) combined with ultrasound homogenization has shown to obtain smaller ice crystals and prevent recrystallization [28]. The air introduction (overrun) tends to limit the ice crystal growth during ice cream production (Fig. 6.8) [29].

Frozen dough has become increasingly popular due to its convenience. However, the quality of the bread produced from frozen dough decreases during frozen storage. The addition of biopolymers, mostly hydrocolloids, can improve the stability of dough during the freezing-thaw cycles and the baking performance (such as increase water retention and decrease firmness and retrogradation) [30]. For example, gamma-polyglutamic acid (γ-PGA), serving as an effective cryoprotectant, has shown to decrease the freezable water content and inhibit the water redistribution facilitating the formation of small and uniform ice crystals during frozen storage and freeze–thaw cycle of frozen steamed bread dough. Moreover, through electrostatic attractions and hydrogen bonds, γ-PGA also alleviated the glutenin depolymerization and improved the fermentability and viscoelasticity of the frozen dough [31].

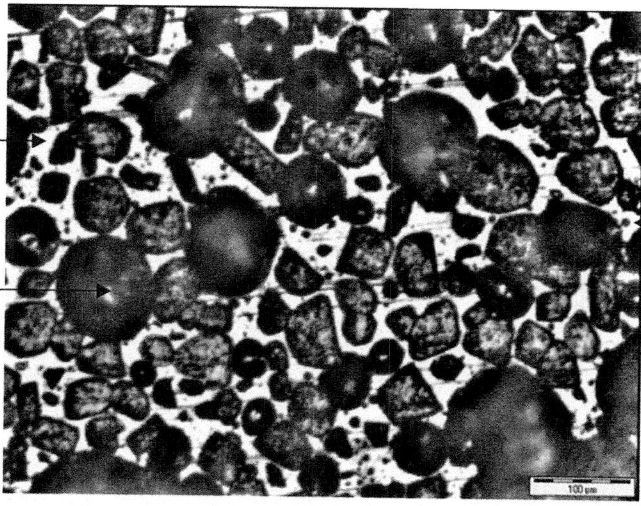

Fig. 6.8 Structure of ice cream observed by direct microscopy method. *Adapted from* [29] *with permission from Elsevier*

Biopolymers have also been applied in food drying including freeze-drying. For example, biopolymers, maltodextrins in particular, have been used as drying aids in spray-drying and freeze-drying of fruit juices and pulps. The benefits of biopolymers in preventing structural changes during drying is thought to be related to their effect of increasing the glass transition temperature (T_g) of natural products [32]. The typical crunchy characteristics of the snack are lost at the onset of glass transition during freeze-drying. Different biopolymers can be added to increase the T_g [33]. For example, gum Arabic and bamboo fiber were added to a grapefruit puree during freeze-drying. The addition did not significantly affect the drying kinetics but resulted in a crunchier and firmer product [34]. Recently, a dehydration film, sodium carboxymethyl cellulose-gum Arabic with anthocyanins, was developed for meat drying (based on the capillary tension principle through hydrogen bonding with water molecules). The results showed that the film-dried (with anthocyanins) meats exhibit more stable pH and color, higher rehydration rate, better sensory quality and microbial growth inhibition compared to the control groups: film without anthocyanins and without film (Fig. 6.9) [35].

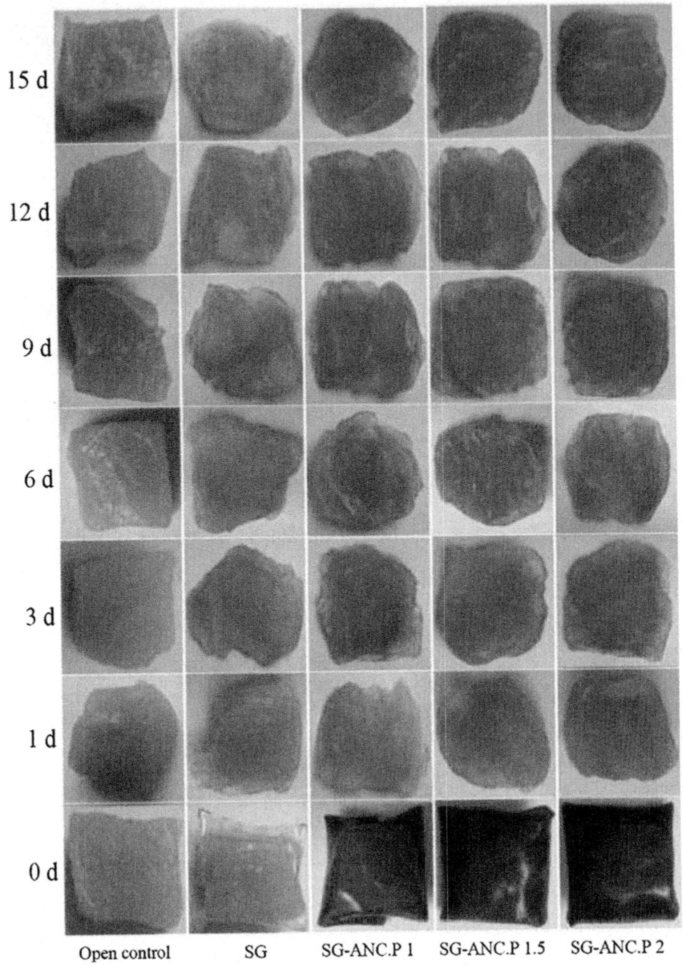

Fig. 6.9 Photos of chicken slices during drying at different time points. SG: sodium carboxymethyl cellulose-gum Arabic, ANC.P: anthocyanins from *Cinnamomum camphora* fruit peel. *Adapted from* [35] *with permission from Elsevier*

References

1. Shi C, Yuan S-C, Zhang L, et al. Coating seeds with biopolymers extracted from waste-activated sludge. Sustainable Chemistry and Pharmacy. 2024;39:101519.
2. Britt DW. Plug-and-play bioinspired seed coatings. Nature Food. 2021;2(7):456–457.
3. Zvinavashe AT, Laurent J, Mhada M, et al. Programmable design of seed coating function induces water-stress tolerance in semi-arid regions. Nature Food. 2021;2(7):485–493.

4. Tomadoni B, Casalongué C, Alvarez VA. Biopolymer-Based Hydrogels for Agriculture Applications: Swelling Behavior and Slow Release of Agrochemicals. In: Gutiérrez TJ, editor. Polymers for Agri-Food Applications. Springer International Publishing; 2019. p. 99–125.
5. Radoslav G, Vukic M, Gojkovic Cvjetkovic V. Application of Biopolymers in the Food Industry. 2017. p. 103–119.
6. McClements DJ. CHAPTER 4 - Biopolymers in Food Emulsions. In: Kasapis S, Norton IT, Ubbink JB, editors. Modern Biopolymer Science. Academic Press; 2009. p. 129–166.
7. Gupta I, Cherwoo L, Bhatia R, et al. Biopolymers: Implications and application in the food industry. Biocatalysis and Agricultural Biotechnology. 2022;46:102534.
8. Katsnelson A. 3D Printers Enter the Kitchen. ACS Central Science. 2021;7(12):1959–1962.
9. Bandoim L. World's First 3D Bioprinted And Cultivated Ribeye Steak Is Revealed. Forbes. 2021. Available from: https://www.forbes.com/sites/lanabanm/2021/02/12/worlds-first-3d-bioprinted-and-cultivated-ribeye-steak-is-revealed/
10. Chen Y, Zhang M, Bhandari B. 3D Printing of Steak-like Foods Based on Textured Soybean Protein. Foods. 2021;10(9):2011.
11. Müller C, Yamashita F, Grossmann M, et al. Films and Coatings Produced from Biopolymers and Composites. 2012. p. 145–216.
12. Khandeparkar AS, Paul R, Sridhar A, et al. Eco-friendly innovations in food packaging: A sustainable revolution. Sustainable Chemistry and Pharmacy. 2024;39:101579.
13. Westlake JR, Tran MW, Jiang Y, et al. Biodegradable biopolymers for active packaging: demand, development and directions. Sustainable Food Technology. 2023;1(1):50–72.
14. Tan W, McClements DJ, Chen J, et al. Novel biopolymer-based active packaging material for food applications: Cinnamaldehyde-loaded calcium nanoparticles incorporated into alginate-carboxymethyl cellulose films. Food Packaging and Shelf Life. 2024;45:101351.
15. Bhowmik S, Agyei D, Ali A. Smart chitosan films as intelligent food packaging: An approach to monitoring food freshness and biomarkers. Food Packaging and Shelf Life. 2024;46:101370.
16. Zhao R, Chen J, Yu S, et al. Active chitosan/gum Arabic-based emulsion films reinforced with thyme oil encapsulating blood orange anthocyanins: Improving multi-functionality. Food Hydrocolloids. 2023;134:108094.
17. Kim DN, Lim J, Bae IY, et al. Effect of hydrocolloid coatings on the heat transfer and oil uptake during frying of potato strips. Journal of Food Engineering. 2011;102(4):317–320.
18. Gonçalves EM, Silva M, Andrade L, et al. From Fields to Films: Exploring Starch from Agriculture Raw Materials for Biopolymers in Sustainable Food Packaging. Agriculture. 2024;14(3):453.
19. Vanaraj R, Suresh Kumar SM, Mayakrishnan G, et al. A Current Trend in Efficient Biopolymer Coatings for Edible Fruits to Enhance Shelf Life. Polymers. 2024;16(18):2639.
20. Pillai ARS, Eapen AS, Zhang W, et al. Polysaccharide-Based Edible Biopolymer-Based Coatings for Fruit Preservation: A Review. Foods. 2024 May 14;13(10):1529.
21. da Mata Cunha O, Lima AMF, Assis OBG, et al. Amphiphilic diethylaminoethyl chitosan of high molecular weight as an edible film. International Journal of Biological Macromolecules. 2020;164:3411–3420.
22. Gupta D, Lall A, Kumar S, et al. Plant-based edible films and coatings for food-packaging applications: recent advances, applications, and trends. Sustainable Food Technology. 2024 2024;2(5):1428–1455.
23. Regand A, Goff HD. Effect of Biopolymers on Structure and Ice Recrystallization in Dynamically Frozen Ice Cream Model Systems. Journal of Dairy Science. 2002;85(11):2722–2732.
24. Hartel RW, Rankin SA, Bradley RL. A 100-Year Review: Milestones in the development of frozen desserts. Journal of Dairy Science. 2017;100(12):10014–10025.
25. Hartel R. Engineering Frozen Desserts. Chemical Engineering Progress. 2019:39–43.

26. Zhu Z, Zhou Q, Sun D-W. Measuring and controlling ice crystallization in frozen foods: A review of recent developments. Trends in Food Science & Technology. 2019;90:13–25.
27. Roos YH. Glass Transition and Re-Crystallization Phenomena of Frozen Materials and Their Effect on Frozen Food Quality. Foods. 2021;10(2):447.
28. Kot A, Jakubczyk E, Kamińska-Dwórznicka A. The Effectiveness of Combination Stabilizers and Ultrasound Homogenization in Milk Ice Cream Production. Applied Sciences. 2023;13(13):7561.
29. Caillet A, Cogné C, Andrieu J, et al. Characterization of ice cream structure by direct optical microscopy. Influence of freezing parameters. LWT—Food Science and Technology. 2003;36(8):743–749.
30. Zaritzky N. Biopolymers Used as Cryoprotectants in Food Freezing. 2012. p. 327–384.
31. Guan E, Zhang T, Wu K, et al. Physicochemical properties and gluten structures of frozen steamed bread dough under freeze–thaw treatment affected by gamma-polyglutamic acid. Food Hydrocolloids. 2023;137:108334.
32. Nicoletti V, Martínez-Navarrete N. Biopolymers used as drying aids in spray and freeze drying of fruit juices and pulps. Biopolymer Engineering in Food Processing. 2012:279–326.
33. Silva-Espinoza MA, Camacho MdM, Martínez-Navarrete N. Use of different biopolymers as carriers for purposes of obtaining a freeze-dried orange snack. LWT. 2020;127:109415.
34. Egas-Astudillo LA, Martínez-Navarrete N, Camacho MM. Impact of biopolymers added to a grapefruit puree and freeze-drying shelf temperature on process time reduction and product quality. Food and Bioproducts Processing. 2020;120:143–150.
35. Alnadari F, Al-Dalali S, Nasiru MM, et al. A new natural drying method for food packaging and preservation using biopolymer-based dehydration film. Food Chemistry. 2023;404:134689.

Biopolymers: Environmental Applications 7

7.1 Introduction to Environmental Applications of Biopolymers

Biopolymers play an essential role in both preventing and addressing environmental issues. From the preventive aspects, biopolymers can be used to produce materials as replacements of conventional materials. For example, bioplastics can be used to replace traditional plastics to reduce white pollution. Bioplastics (manufactured from bio-based polymers) are plastics that are either made from renewable resources (bio-based), are biodegradable, are made through biological processes or a combination of these (biodegradable but fossil-based plastics are not considered bioplastics) [1]. There are still various challenges for the wider application of bioplastics: economic feasibility, industrial-scale implementation, usage of additives during bioplastic production, agricultural impacts, unclear end-of-life management, and biodegradability assessment [1–3] (Fig. 7.1).

Biopolymers have been widely used in treatment of wastewater from different sources (Sect. 7.2). Moreover, they have also been commonly applied in bioremediation and soil stabilization (Sect. 7.3). There are many other environmental applications of biopolymers such as scale inhibition and electronic waste treatment. For example, biopolymers have been as scale inhibitors in the petroleum industry. Mineral scale formation and deposition occurs during certain operations in petroleum production. Some of the biodegradable polymers used as scale inhibitors include poly(aspartic acid), poly(epoxysuccinic acid), carboxymethyl inulin, carboxymethyl chitosan and carboxymethyl cellulose [4]. Crosslinked biopolymer-based 3D structures (e.g., bioconcrete) can serve as effective tools to stabilize such hazardous materials as heavy metals, toxic organics, and radioactive waste in a subsurface engineering system [5–7]. The structure of polymeric concrete is shown in Fig. 7.2. A concrete crosslinked with biopolymers, xanthan gum and guar gum

W. Zhang, *Biopolymer Engineering*, Synthesis Lectures on Chemical Engineering and Biochemical Engineering, https://doi.org/10.1007/978-3-031-98868-4_7

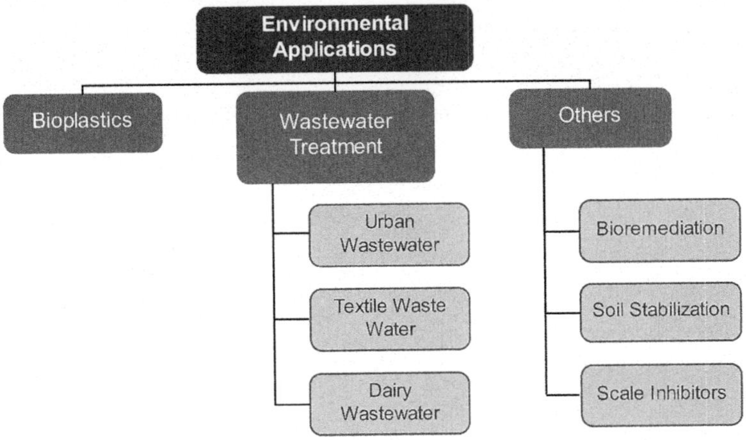

Fig. 7.1 Environmental applications of biopolymers

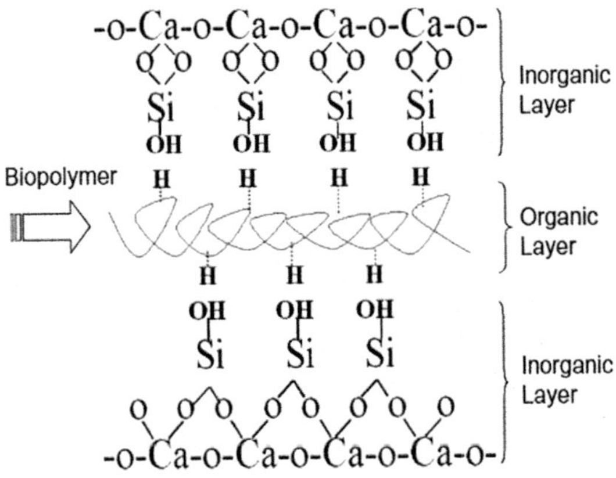

Fig. 7.2 Structure of biopolymeric concrete. *Adapted from [9] with permission from Taylor and Francis*

(Fig. 7.3), was used to encapsulate cathode ray tubes (CRT) waste, reducing Pb leachability [8]. Cross-linked biopolymeric concrete, based on chitosan or xanthan gum/guar gum, was studied for selenium glass wastes [9]. Xanthan and guar gum mix provides the synergistic effect of connecting xanthan gum to unsubstituted mannose regions of guar gum, and subsequently binding metal ions (due to the functions groups: carboxyl (xanthan gum) and hydroxyl (guar gum)) onto the concrete structure (Fig. 7.3) [8].

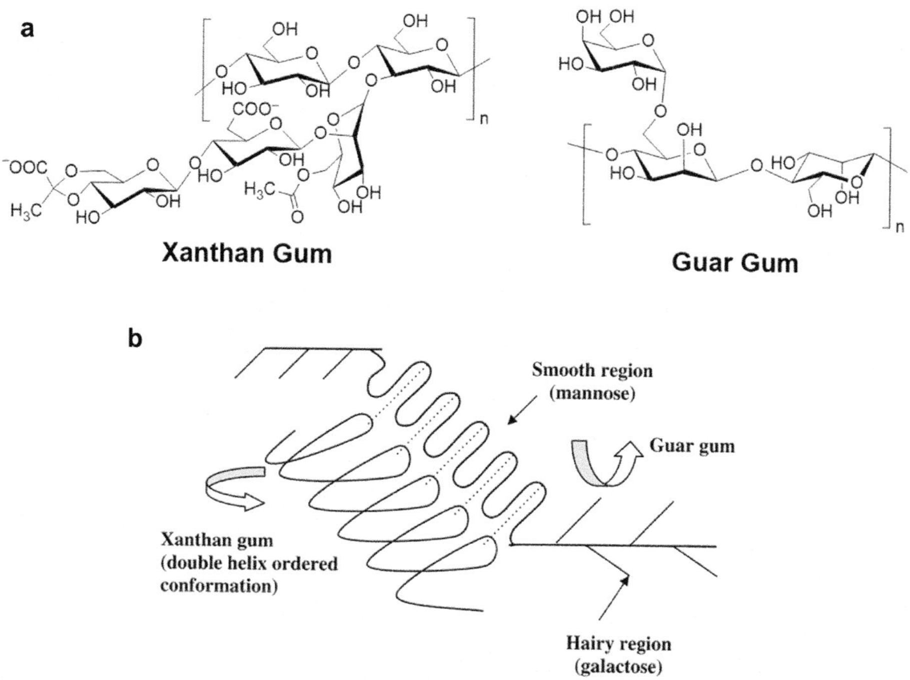

Fig. 7.3 Structure of xanthan and guar gum (**a**) and synergistic interactions between xanthan and guar gum (**b**). *(b) is adapted from [8] with permission from Elsevier*

7.2 Biopolymers in Wastewater Treatment

Biopolymers have been used in various wastewater treatment, such as urban, textile, and dairy wastewater. Both biochemical oxygen demand (BOD) and chemical oxygen demand (COD) are important measures of water quality and are routine analyses for wastewater [10]. BOD measures the amount of dissolved oxygen needed by microbes to breakdown organic matter present in a given water sample and is commonly expressed in mg of oxygen consumed per L sample during 5 days of incubation at 20 °C. COD measures all substances that can be chemically oxidized (not just biodegradable organic matter; hence, generally higher than BOD) [11, 12]. COD yields results much faster compared to BOD [13]. Higher BOD/COD indicates lower water quality. Coagulation and flocculation additives are used for clarification of urban waters.

Coagulants separate suspended particles (colloids) from water by destabilizing them leading to the formation of small aggregates of particles (through various mechanisms such as charge neuralization, polymer adsorption, interparticle bridging, and double layer compression; Fig. 7.4). Aggregated particles then form settleable flocs (i.e., flocculation) followed by sedimentation, filtration and disinfection [14–17]. Biopolymers, such as alginate, cellulose, and chitosan-based materials, offer a more sustainable and eco-friendly option for removing contaminants from wastewater without introducing harmful chemicals and/or byproduct compared conventional chemical treatments [18].

Biopolymers with functional groups, such as amine, hydroxide, phenolic hydroxyl, and methoxy carboxyl, have been used as adsorbents of pollutants in water. Some of the promising biopolymers for commercial applications include cellulose (and its

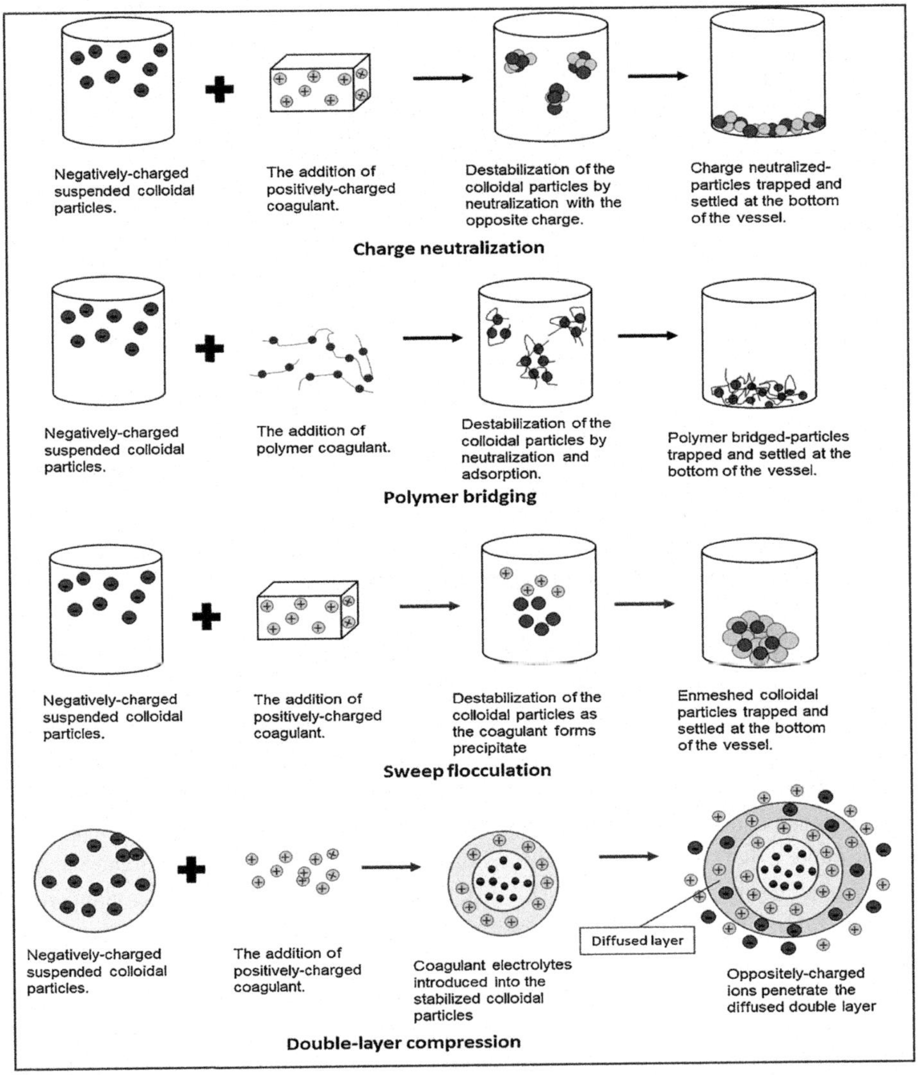

Fig. 7.4 Schematic of different coagulation mechanisms. *Adapted from [17] with permission from Elsevier*

derivatives), chitosan (and its derivatives) and poly(glutamic acid) [19, 20]. In particular, cellulose-based materials are effective for removing heavy metals, dyes, pathogenic microorganisms, and other pollutants from wastewater [20]. Although promising, the full environmental impact of using biopolymers in wastewater treatment should be carefully considered to assess their sustainability and compatibility. Some of the concerns include contaminant release, additive leaching, and environmental footprint [21].

Certain types of wastewater, such as textile and dairy wastewater, can be challenging for treatment. Textile wastewater has high concentration of organic compounds (high BOD/COD) and treatment is necessary for drinking and other domestic uses [22, 23]. Chemicals present in textile wastewater are dyes, starches, hydrogen peroxide, surfactants, acids, alkalis, metal soaps, and dispersants [24]. The dye adsorption mechanism of biopolymers is dye type-dependent and are predominantly through dipole-dipole, hydrogen bonding, π-π, or electrostatic interactions [25]. Chitosan-based materials have been investigated for dye removal—due to its ability to form linkages with the sulfonic groups of anionic dyes. Chitosan modifications through crosslinking, graft copolymerization and surfactant impregnation have been studied to improve its dye sorption properties [26]. Additionally, decrystallized chitosan with a low degree of crystallinity was shown to be more efficient in decolorization compared to normal chitosan and activated carbon which might due to the high penetration of dyes into its structure leading to higher binding capacity [27]. Cellulose-based adsorbents are also attractive due to their regeneratable and recyclable properties. Chemical modification of cellulose, creating more active binding sites, has been widely studied to improve its adsorption capacity [28]. Alginate beads, prepared using different types of cross-liking, have also been studied for dye removal. The dry epichlorohydrin cross-linked alginate beads (DEpAB) showed the highest efficiency (Fig. 7.5) and the extend of adsorption was affected by various factors, such as pH [29].

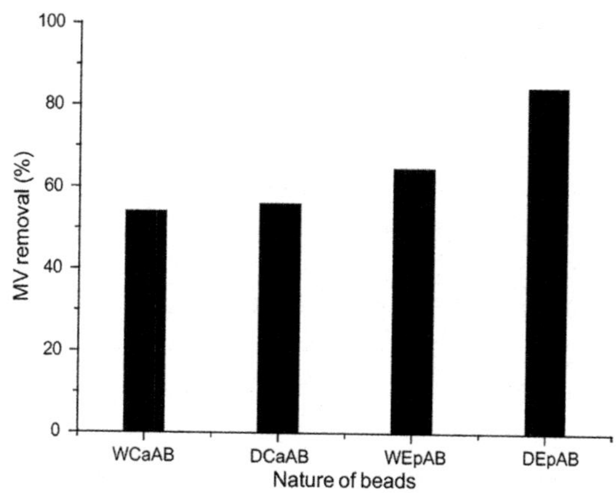

Fig. 7.5 Effect of the nature of beads on the MV percentage removal. MV: methyl violet, W: wet, D: dry, AB: alginate beads, Ep: epichlorohydrin. *Adapted from [29] with permission from Springer Nature*

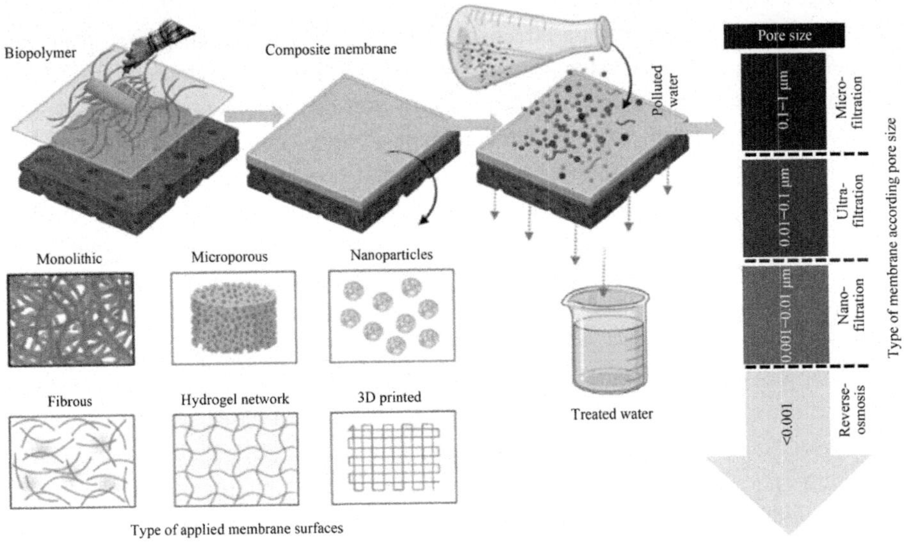

Fig. 7.6 Types of applied membrane surfaces along with their use according to pore sizes. *Adapted from [33] with permission from Elsevier*

Dairy wastewater is of high organic and nutrient loading and extreme pH variations [30]. Carboxymethyl cellulose is commonly used for dairy waste treatment after reducing the pH to its pI. Chitosan, produced from shrimp-shell waste, was shown to treat dairy wastewater with higher efficiencies and at higher pH values [31]. Membrane filtration methods, in particular microfiltration, ultrafiltration, nanofiltration, and reverse osmosis show great potential for dairy wastewater treatment (Fig. 7.6) [32, 33]. Biopolymers, such as cellulose and poly(lactic acid), have been explored for membrane synthesis [33].

Activated-sludge has widely been used for wastewater treatment, where air is incorporated into a bioreactor with wastewater and microorganisms allowing the growth of heterotrophic and autotrophic bacteria to metabolize organic matter and nutrients [34]. Activated-sludge and microorganisms encapsulation with biopolymers, such as poly(vinyl alcohol) and alginate have been performed to increase the nutrients from wastewater [34–37]. Polymeric biocarriers (biofilm carriers) have been studied to allow the attachment of microbes during wastewater treatment [38]. For example, multifunctional modified poly(vinyl alcohol) carriers were developed to enhance simultaneous denitrification, Cd(II) and Mn(II) removal efficiency in bioreactor environments [39] (Fig. 7.7).

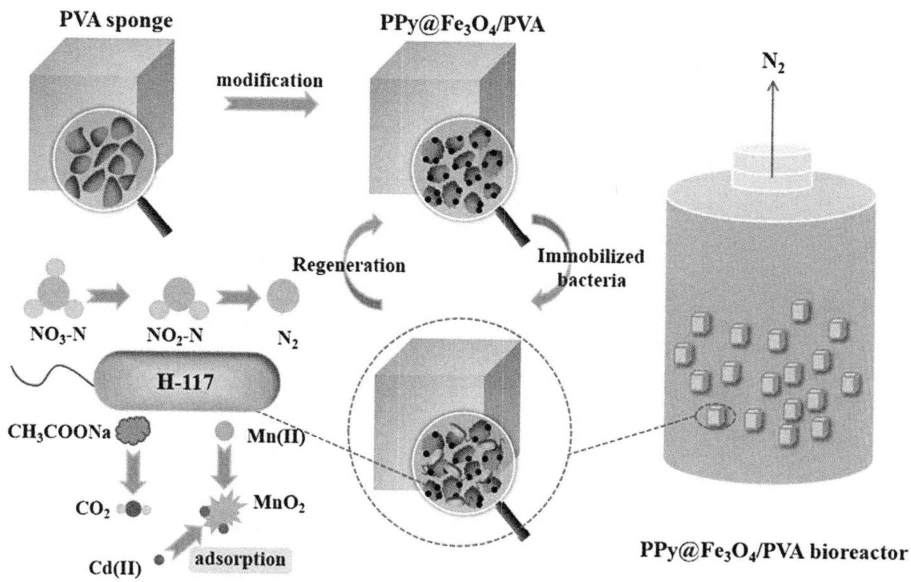

Fig. 7.7 Multifunctional modified polyvinyl alcohol (PVA) bacterial carrier for enhancing bioreactor performance in nitrate, Mn(II) and Cd(II) removal. PPy: polypyrrole, H-117: Pseudomonas sp. H117. *Adapted from [39] with permission from Elsevier*

7.3 Biopolymers in Bioremediation and Soil Stabilization

Bioremediation refers to the use of microorganisms to eliminate or reduce the concentrations of hazardous wastes at a contaminated site [40]. It is an approach usually employed for degrading, neutralizing, and detoxifying contaminants predominantly from soil, water, and sediments—performed by diverse means including biodegradation, phytoremediation, bioaugmentation, and biostimulation [41]. The definition has been expanded to include genetically modified organisms and biomaterials that mimic biological processes [42]. Biomimetic multifunctional lignocellulosic nanoframes were developed for in situ bioremediation via fungus *Irpex lacteus* and promote per- and polyfluoroalkyl substances detoxification (Fig. 7.8) [43]. Encapsulation of microorganisms has also been extensively studied for bioremediation. Commonly used biopolymers for encapsulation include alginate, chitosan, and poly(vinyl alcohol). Encapsulation serves two functions: protection of exogenous microorganisms and controlled release [44]. Interestingly, 3D bioprinting has also been investigated for bioremediation offering scalable, customizable, and potentially greener solutions [42].

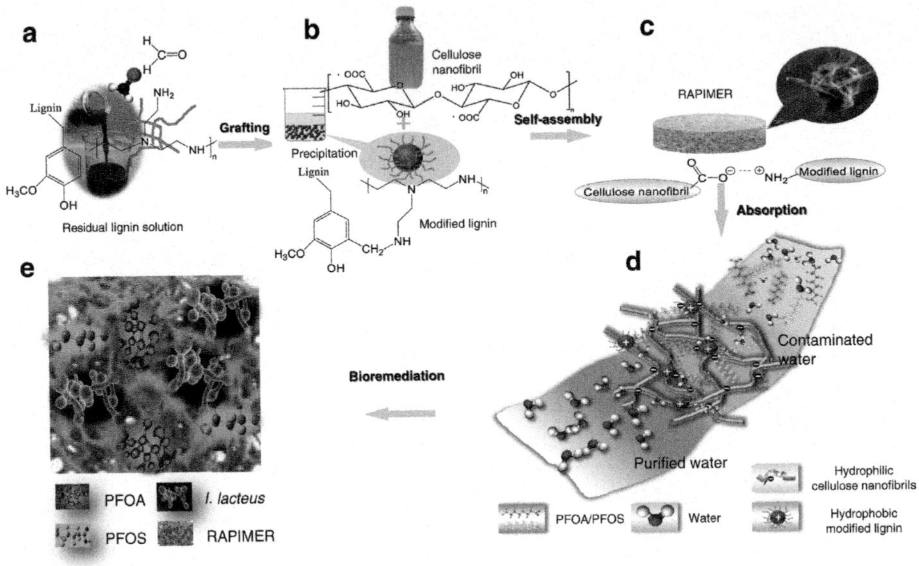

Fig. 7.8 The design strategy, fabrication process, chemical adsorption, and fungus degradation scheme of the RAPIMER system. RAPIMER: renewable artificial plant for In-situ microbial environmental remediation, PFOA: perfluorooctanoic acid, PFOS: perfluorooctanesulfonic acid. *Adapted from [43]*

Soil stabilization is an approach for altering the physiochemical attributes of the soil. It enhances the soil's effectiveness in terms of physical aspects (improve the structural stability of soil). Different biopolymers, such as xanthan gum, guar gum, gellan gum, polyacrylamide, and chitosan, have been used for dust control, erosion control, and strengthening and stabilizing soil [41, 45–47]. For example, it was demonstrated that both xanthan gum and carboxymethyl cellulose, when applied individually, were effective stabilizers of mine residue soil [48]. Another study showed that the xanthan gum treatment gave the soils stronger cementation and a denser structure increasing strength and brittleness [49] (Fig. 7.9). Different types of biopolymers, xanthan gum, guar gum, β-1,3/1,6-glucan, chitosan, and alginate, were compared, and the results indicated that xanthan gum, guar gum, and β-1,3/1,6-glucan showed the most potential for the enhancement of soil mechanical properties [50].

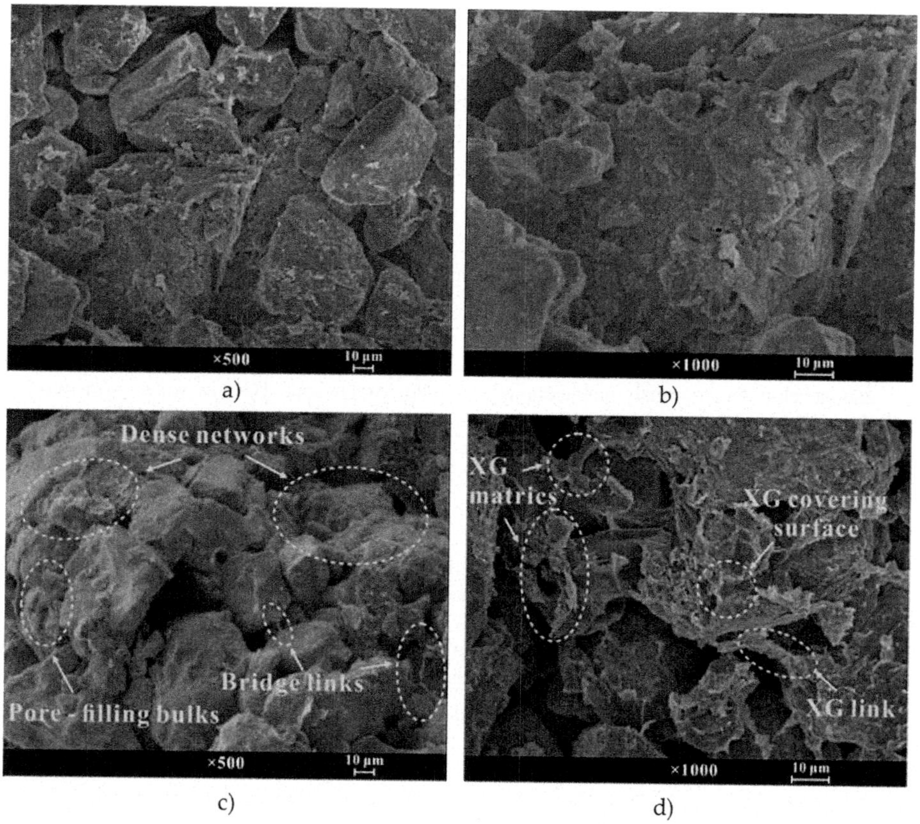

Fig. 7.9 Microstructure of xanthan gum (XG)-treated loess (**a**) and (**b**) SEM image of untreated soil; (**c**) and (**d**) SEM image of the 2.0% XG-treated soil at × 1000. *Adapted from* [49]

References

1. Rosenboom J-G, Langer R, Traverso G. Bioplastics for a circular economy. Nature Reviews Materials. 2022;7(2):117-137.
2. Onen Cinar S, Chong ZK, Kucuker MA, et al. Bioplastic Production from Microalgae: A Review. Int J Environ Res Public Health. 2020;17(11).
3. Folino A, Pangallo D, Calabrò PS. Assessing bioplastics biodegradability by standard and research methods: Current trends and open issues. Journal of Environmental Chemical Engineering. 2023;11(2):109424.
4. Husna UZ, Elraies KA, Shuhili JABM, et al. A review: the utilization potency of biopolymer as an eco-friendly scale inhibitors. Journal of Petroleum Exploration and Production Technology. 2022;12(4):1075-1094.
5. Kim D, Park J-S, Wang R, et al. Studies of crosslinked biopolymer structure for environmental tools in terms of the rheological measurements: part B. Environmental Earth Sciences.

2021;80(8):311.
6. Yao Z, Ling T-C, Sarker PK, et al. Recycling difficult-to-treat e-waste cathode-ray-tube glass as construction and building materials: A critical review. Renewable and Sustainable Energy Reviews. 2018;81:595-604.
7. Soleimani M, Shahandashti M. Comparative process-based life-cycle assessment of bioconcrete and conventional concrete. Journal of Engineering, Design and Technology. 2017;15(5):667-688.
8. Kim D, Quinlan M, Yen TF. Encapsulation of lead from hazardous CRT glass wastes using biopolymer cross-linked concrete systems. Waste Management. 2009;29(1):321-328.
9. Kim D, Park J-S, Yen TF. Feasibility study on cross-linked biopolymeric concrete encapsulating selenium glass wastes. Journal of the Air & Waste Management Association. 2012;62(8):898-904.
10. Aguilar-Torrejón JA, Balderas-Hernández P, Roa-Morales G, et al. Relationship, importance, and development of analytical techniques: COD, BOD, and, TOC in water—An overview through time. SN Applied Sciences. 2023;5(4):118.
11. Hlordzi V, Kuebutornye FKA, Afriyie G, et al. The use of Bacillus species in maintenance of water quality in aquaculture: A review. Aquaculture Reports. 2020;18:100503.
12. Li D, Liu S. Chapter 12 - Water Quality Monitoring in Aquaculture. In: Li D, Liu S, editors. Water Quality Monitoring and Management: Academic Press; 2019. p. 303–328.
13. Hu Z, Grasso D. WATER ANALYSIS | Chemical Oxygen Demand. In: Worsfold P, Townshend A, Poole C, editors. Encyclopedia of Analytical Science (Second Edition). Oxford: Elsevier; 2005. p. 325-330.
14. Saritha V, Srinivas N, Srikanth Vuppala NV. Analysis and optimization of coagulation and flocculation process. Applied Water Science. 2017;7(1):451-460.
15. Muruganandam L, Saravana Kumar MP, Jena A, et al. Treatment of waste water by coagulation and flocculation using biomaterials. IOP Conference Series: Materials Science and Engineering. 2017;263(3):032006.
16. Suopajärvi T. Functionalized Nanocelluloses in Wastewater Treatment Applications 2015.
17. Owodunni AA, Ismail S. Revolutionary technique for sustainable plant-based green coagulants in industrial wastewater treatment–A review. Journal of Water Process Engineering. 2021;42:102096.
18. Saiyad M, Shah N, Joshipura M, et al. Modified biopolymers in wastewater treatment: A review. Materials Today: Proceedings. 2024.
19. Kolya H, Kang C-W. Bio-Based Polymeric Flocculants and Adsorbents for Wastewater Treatment. Sustainability. 2023;15(12):9844.
20. Sharma R, Nath PC, Mohanta YK, et al. Recent advances in cellulose-based sustainable materials for wastewater treatment: An overview. International Journal of Biological Macromolecules. 2024;256:128517.
21. Kasbaji M, Mennani M, Oubenali M, et al. Bio-based functionalized adsorptive polymers for sustainable water decontamination: A systematic review of challenges and real-world implementation. Environmental Pollution. 2023;335:122349.
22. Senthil Kumar P, Saravanan A. 11 - Sustainable wastewater treatments in textile sector. In: Muthu SS, editor. Sustainable Fibres and Textiles: Woodhead Publishing; 2017. p. 323–346.
23. Holkar CR, Jadhav AJ, Pinjari DV, et al. A critical review on textile wastewater treatments: Possible approaches. Journal of Environmental Management. 2016;182:351-366.
24. Navidjouy N, Soltani F, Rahimnejad M. Chapter 12 - Application of biological fuel cell in wastewater treatment. In: Rahimnejad M, editor. Biological Fuel Cells: Elsevier; 2023. p. 301–320.

25. Dassanayake RS, Acharya S, Abidi N. Recent Advances in Biopolymer-Based Dye Removal Technologies. Molecules. 2021;26(15).
26. Qamar SA, Ashiq M, Jahangeer M, et al. Chitosan-based hybrid materials as adsorbents for textile dyes–A review. Case Studies in Chemical and Environmental Engineering. 2020;2:100021.
27. Trung TS, Ng C-H, Stevens WF. Characterization of decrystallized chitosan and its application in biosorption of textile dyes. Biotechnology Letters. 2003;25(14):1185-1190.
28. Kausar A, Zohra ST, Ijaz S, et al. Cellulose-based materials and their adsorptive removal efficiency for dyes: A review. International Journal of Biological Macromolecules. 2023;224:1337-1355.
29. Merakchi A, Bettayeb S, Drouiche N, et al. Cross-linking and modification of sodium alginate biopolymer for dye removal in aqueous solution. Polymer Bulletin. 2019;76(7):3535-3554.
30. Slavov AK. General Characteristics and Treatment Possibilities of Dairy Wastewater - A Review. Food Technol Biotechnol. 2017;55(1):14-28.
31. Selmer-Olsen E, Ratnaweera HC, Pehrson R. A novel treatment process for dairy wastewater with chitosan produced from shrimp-shell waste. Water Science and Technology. 1996;34(11):33-40.
32. Al-Tayawi AN, Sisay EJ, Beszédes S, et al. Wastewater Treatment in the Dairy Industry from Classical Treatment to Promising Technologies: An Overview. Processes. 2023;11(7):2133.
33. Elgarahy AM, Eloffy MG, Guibal E, et al. Use of biopolymers in wastewater treatment: A brief review of current trends and prospects. Chinese Journal of Chemical Engineering. 2023;64:292-320.
34. Guerrero JA, Almeida-Naranjo CE, Villamar Ayala CA. Improvement of nutrients removal from domestic wastewater by activated-sludge encapsulation with polyvinyl alcohol (PVA). Journal of Environmental Science and Health, Part A. 2019;54(8):721-727.
35. Han Y, Zhang W, Lu W, et al. Co-immobilization of Pseudomonas stutzeri YHA-13 and Alcaligenes sp. ZGED-12 with polyvinyl alcohol–alginate for removal of nitrogen and phosphorus from synthetic wastewater. Environmental Technology. 2014;35(22):2813–2820.
36. Deng J, Chen Q, Hu B, et al. Synergic Effect of Adsorption and Biodegradation by Microsphere Immobilizing Bacillus velezensis for Enhanced Removal Organics in Slaughter Wastewater. Processes. 2021;9(7):1145.
37. Yadav P, Yadav A, Srivastava JK, et al. Reduction of pollution load of tannery effluent by cell immobilization approach using Ochrobactrum intermedium. Journal of Water Process Engineering. 2021;41:102059.
38. Al-Amshawee S, Yunus MYBM, Vo D-VN, et al. Biocarriers for biofilm immobilization in wastewater treatments: a review. Environmental Chemistry Letters. 2020;18(6):1925–1945.
39. Su Jf, Bai Yh, Huang Tl, et al. Multifunctional modified polyvinyl alcohol: A powerful biomaterial for enhancing bioreactor performance in nitrate, Mn(II) and Cd(II) removal. Water Research. 2020;168:115152.
40. Wu Y, Xia L, Yu Z, et al. In situ bioremediation of surface waters by periphytons. Bioresource Technology. 2014;151:367-372.
41. Sharma M, Tellili N, Kacem I, et al. Microbial Biopolymers: From Production to Environmental Applications–A Review. Applied Sciences. 2024;14(12):5081.
42. Finny AS. 3D bioprinting in bioremediation: a comprehensive review of principles, applications, and future directions. PeerJ. 2024;12.
43. Li J, Li X, Da Y, et al. Sustainable environmental remediation via biomimetic multifunctional lignocellulosic nano-framework. Nature Communications. 2022;13(1):4368.
44. Valdivia-Rivera S, Ayora-Talavera T, Lizardi-Jiménez MA, et al. Encapsulation of microorganisms for bioremediation: Techniques and carriers. Reviews in Environmental Science and Bio/Technology. 2021;20(3):815-838.

45. Chang I, Lee M, Tran ATP, et al. Review on biopolymer-based soil treatment (BPST) technology in geotechnical engineering practices. Transportation Geotechnics. 2020;24:100385.
46. Zhang J, Liu J. A Review on Soils Treated with Biopolymers Based on Unsaturated Soil Theory. Polymers. 2023;15(22):4431.
47. Jang J. A Review of the Application of Biopolymers on Geotechnical Engineering and the Strengthening Mechanisms between Typical Biopolymers and Soils. Advances in Materials Science and Engineering. 2020;2020(1):1465709.
48. Correia AAS, Caldeira JB, Branco R, et al. Enhancing the Strength of Mine Residue Soil by Bioremediation Combined with Biopolymers. Applied Sciences. 2023;13(18):10550.
49. Jiang T, Zhao J-d, Zhang J-r. Splitting tensile strength and microstructure of xanthan gum-treated loess. Scientific Reports. 2022;12(1):9921.
50. Soldo A, Miletić M, Auad ML. Biopolymers as a sustainable solution for the enhancement of soil mechanical properties. Scientific Reports. 2020;10(1):267.

Engineering Biopolymers: Market and Regulations

8

8.1 Market of Biopolymers

The biopolymer market is expected to grow explosively. For example, the market of biopolymers produced through natural resources is anticipated to have a 22.4% annual growth from \$10.7 billion in 2022 to an estimated \$86.7 billion by 2032 [1]. The worldwide bio-based polymer production capacity is expected to increase to 9.3 million tons in 2027 from 4.9 million tons in 2022 (about 14% compound annual growth rate (CAGR)). Asia is the leading region in bio-based polymer production with the highest growth capacities between 2022 and 2027. Among the biopolymers, poly(lactic acid) (PLA) and polyhydroxyalkanoates (PHAs) are expected to have the largest growth [2]. Some of the companies in producing biopolymer and/or polymer-based product are listed in Table 8.1.

Biopolymers are commonly used to produce bioplastics. Bioplastics that are plastics that are either made from renewable resources (bio-based), are biodegradable, are made through biological processes or a combination of these. However, it is not recommended to use this terminology for plastics that are biodegradable but fossil-based. Bio-based plastics are not necessary more sustainable than fossil-based plastics but can reduce the carbon emissions due to the use of renewable resources [3].

When it comes to biopolymer engineering, material properties, such as mechanical, thermal, and processing, are critical [4]. One of the classic books for this topic in the market is Engineering Biopolymer: Markets, Manufacturing, Properties and Applications [5]. Regarding the mechanical properties, biopolymers cover a broad range of stiffness and impact toughness properties, comparable to conventional polymers. Among common biopolymers, cellulose derivates and PLA show the highest modulus of elasticity (Fig. 8.1). However, biopolymers exhibit lower heat resistance compared to conventional polymers such as polyethylene terephthalate and Nylon 6 (Fig. 8.1). Besides high moisture absorption values of starch blends and cellulose derivates (Fig. 8.1), some biopolymers

W. Zhang, *Biopolymer Engineering*, Synthesis Lectures on Chemical Engineering and Biochemical Engineering, https://doi.org/10.1007/978-3-031-98868-4_8

Table 8.1 Some of the biopolymer companies

Company	Product(s)
BASF	Ecoflex®, Ecovio®, etc
Biome Bioplastics	bio-based polyesters
Biotec	biopolymer compounds made from plant-based renewable resources
Braskem	I'm Green™ biopolymers
Danimer scientific	polyhydroxyalkanoate
Eastman chemical company	Trēva™ engineering bioplastic (cellulose-based), Naia™ cellulosic fiber, etc
Empower materials	CO_2-based polymers
Ecovative	Mushroom® packaging
FlexSea	biopolymer derived from seaweed
Mitsubishi chemical	Polybutylene succinate, DURABIO™ (bio-based polycarbonate, etc
Nature works	poly(lactic acid)
Newlight technologies	AirCarbon (polyhydroxybutyrate)
Novamont	MATER-BI bioplastics
Novomer	poly(3-hydroxypropionate)
Paques biomaterials	polyhydroxyalkanoate
Plantic technologies	PLANTIC™HP, etc
Rodenburg biopolymers	Solanyl® products (potato starch-based)
SABIC	bio-based ULTEM™ resins (polyetherimide)
Teknor apex	Terraloy® PLA, biodegradable co-polyester, etc
Total energies corbion	poly(lactic acid)
Zhejiang hisun biomaterials	poly(lactic acid)

* This is by no means an exhaustive list (in alphabetical order of the company name) and is subjected to updates and additions

are susceptible to certain manufacturing conditions, such as high temperature and high humidity levels. There are other challenges for implementing bioplastics, which include economics (more expensive than fossil-based plastics), efficiency (can be less efficient during manufacturing), end of life (recycling challenges), ethics (e.g., competition with food production) and education [3].

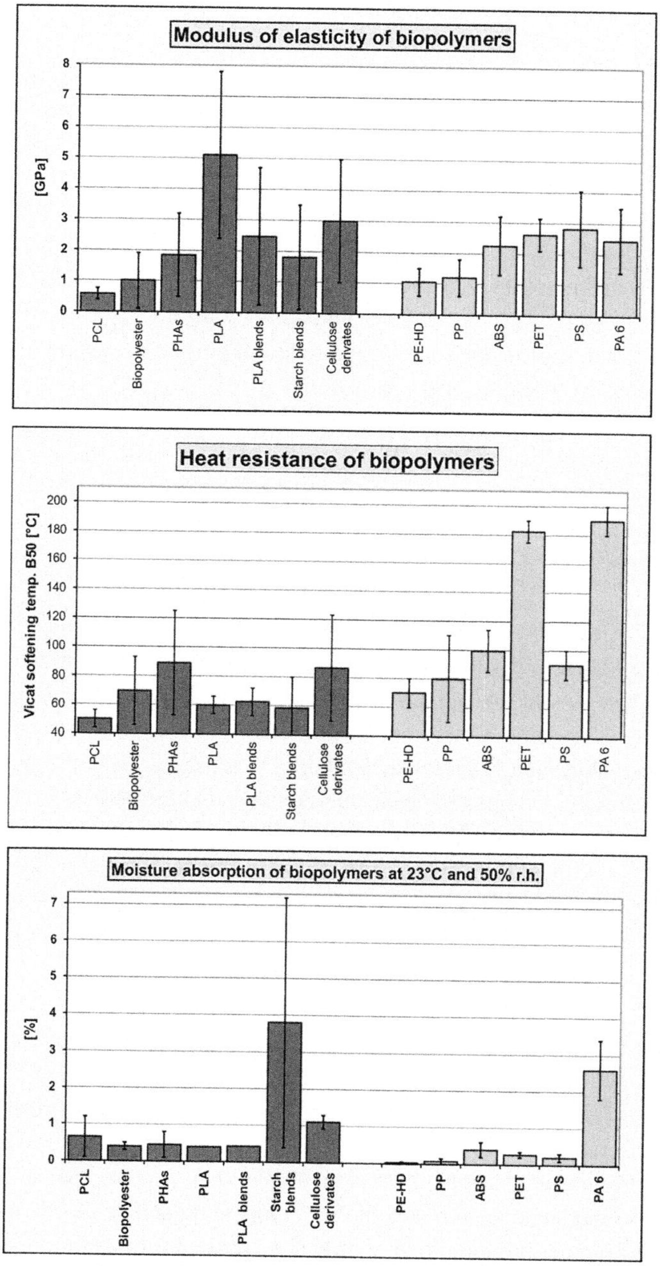

Fig. 8.1 Properties comparison between biopolymers and conventional polymers. PCL: polycaprolactone, PHAs: polyhydroxyalkanoates, PLA: polylactide, PE-HD: polyethylene-high density, PP: polypropylene, ABS: polyacrylonitrile-butadiene-styrene, PET: polyethylene terephthalate, PS: polystyrene, PA 6: polyamide (Nylon 6). *Adapted from* [4] *with permission from Elsevier*

8.2 Regulations of Biopolymers

After using, there are different options for polymer/biopolymer-based products—landfill, recycling, compositing, incineration and more [6]. Plastic pollution, including microplastics remains a major concern and challenge worldwide. Microplastics are defined as solid plastic particles ≤ 5 mm in size, composed of polymers together with functional additives as well as other intentionally and unintentionally added chemicals [7]. Microplastics are ubiquitous not just in the ocean but also on land and in freshwater systems and have exposed significant environment and ecological impacts and risks, including human health (Fig. 8.2) [7–10]. Microplastics can enter human body through inhalation, ingestion, and dermal absorption. Exogenous chemicals, proteins, and toxins present in or on the particle (acting as vectors) may lead to chemical toxicity [7, 8].

Regarding recycling, there are codes based on plastic categories (Fig. 8.3). The code 7 (others) category includes bisphenol A, polycarbonate, and bio-based plastics. Generally, the higher the code number, the more difficult it is for the material to be recycled [11]. As discussed in Sect. 8.1, recycling of biopolymer-based products can be challenging, for example, consumers remain uncertain of how to deal with them after use and the recycling streams haven't been fully established [3].

Biodegradable polymers/biopolymers can be treated either by industrial compositing or digestion (or biogasfication). Industrial composting is the controlled biological decomposition of organic waste under managed conditions that are predominately aerobic and that allow the development of thermophilic conditions as a result of biologically produced exothermic heat. Digestion is anerobic and biogas is generated from the organic substance during the process [6]. Composting generates compost and the process is elevated with the increase of temperature. Anaerobic digestion generates biogas (CH_4 and CO_2) and digestate with little or no heat released [6, 12]. The term biodegradable is not an appropriate marketing term or claim for describing end of life behavior because it lacks specificity on timeframe and environment (compared to compostable), although it is accurate when used in technical contexts [13].

There are a few common certifications for compostable plastics worldwide—Biodegradable Products Institute (BPI, North America), Australasian Bioplastics Association (ABA, Australia and New Zealand), and European Bioplastics e.V. [14]. BPI is the leading third-party verification of ASTM (American Society for Testing and Materials) standards (such as ASTM-D6400) for compostable products in North America [15]. The ABA launched Australian Standard 4736–2006, compostable and biodegradable plastics—"Biodegradable plastics suitable for composting and other microbial treatment" which is known as the 'seedling logo' certification system throughout Australia and New Zealand [16]. In Europe, the harmonised European standard EN 13,432 "Requirements for packaging recoverable through compositing and biodegradation" requires at least 90% disintegration after twelve weeks, 90% biodegradation (CO_2 evolvement) in six months, and include tests on ecotoxicity and heavy metal content. EN 13,432 is for biodegradable

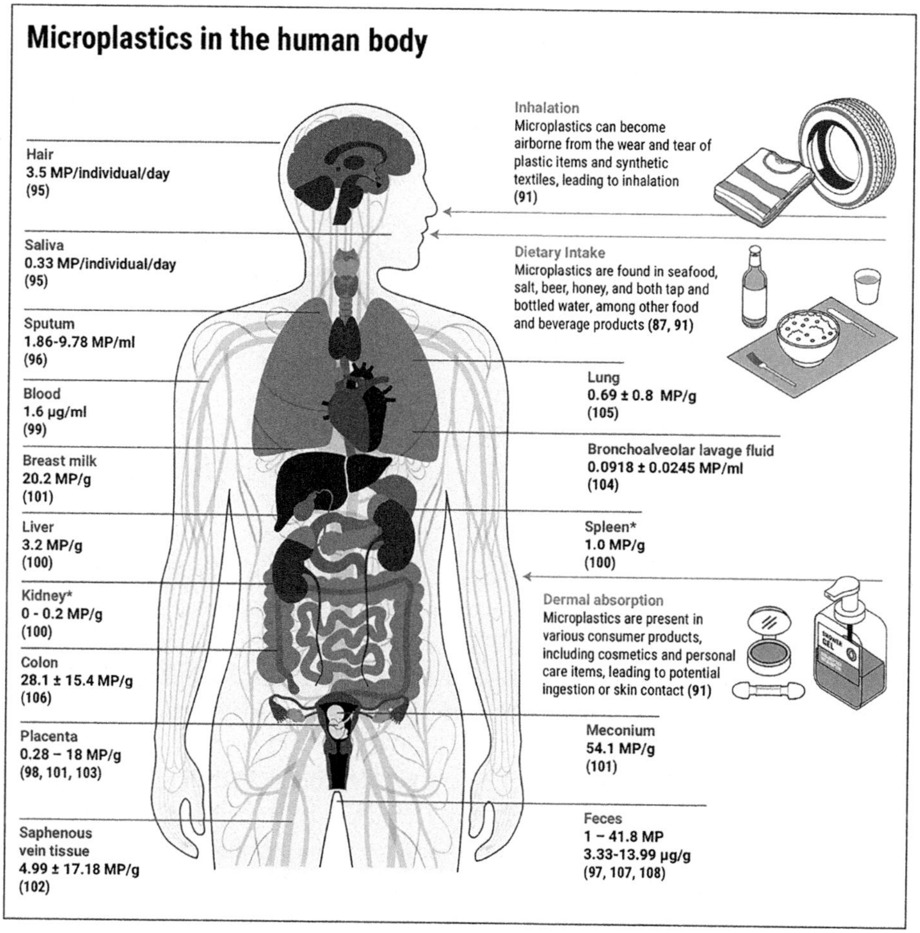

Fig. 8.2 Locations in the human body where microplastics have been reported. *Adaptedfrom* [7] *with permission from the American Association for the Advancement of Science*

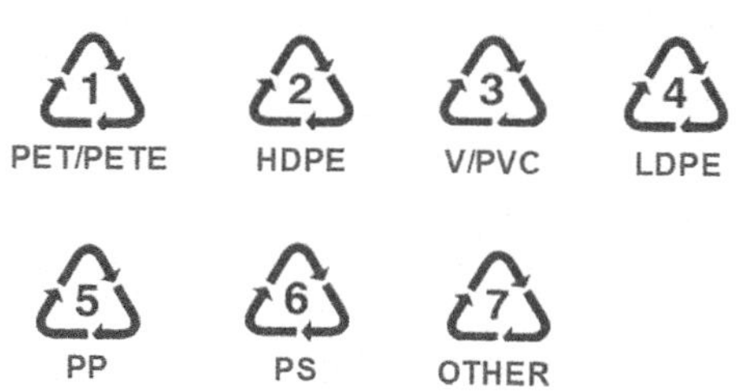

Fig. 8.3 Plastic recycling codes. PET/PETE: poly(ethylene terephthalate), HDPE: high-density polyethylene, V/PVC: poly(vinyl chloride), LDPE: low-density polyethylene, PP: polypropylene, PS: polystyrene

Fig. 8.4 Compostable logos

packaging designed for treatment in industrial composting facilities and anaerobic digestion. EN 14,995, describing the same requirements and tests as EN 13,432, applies not only to packaging but plastics in general. The same holds for ISO (International Standard Organization) 18,606 "Packaging and the environment" and ISO 17088 "Specifications for compostable plastics" [17]. In Asia, the Japan BioPlastics Association (JBPA) also provides certification [18]. To summarize, ISO 17088, EN 13,432, and EN 14,995 determine specific procedures and requirements for the identification and labeling of compostable plastics and products made of compostable plastic materials (strongly relates to the ASTM D6400). ASTM D6400 defines specification for the labeling of plastics that are originally designed to be aerobically composted [19] (Fig. 8.4).

The adoption of circular economy practices is growing within industry and the global society. Circular economy is an economy that is restorative and regenerative by intention (with systemic innovation at the core). It is a new way of creating value, through extending product lifetime and relocating waste from the end of the value chain to the beginning—in effect, using products and their resources more efficiently by using them more than once [20]. The plastics industry has traditionally been based on linear life cycles. In a circular economy, waste becomes raw material for a recycling process at its end of life, and all production and recycling processes are supplied with renewable energy (Fig. 8.5) [3]. Life cycle assessment (LCA) has become widely recognized as an effective tool for assessing the resource use, environmental burdens, and human health impacts connected with the complete life cycle of products, processes, and activities (a holistic technique to identify and quantify the potential impacts of a product or process throughout its life cycle). The ISO 14040 series of standards provides the framework for an LCA, which consists of four interrelated phases: define the goal and scope; compile a life cycle inventory, evaluate the life cycle impacts, and interpreting the results [21]. There are also specific

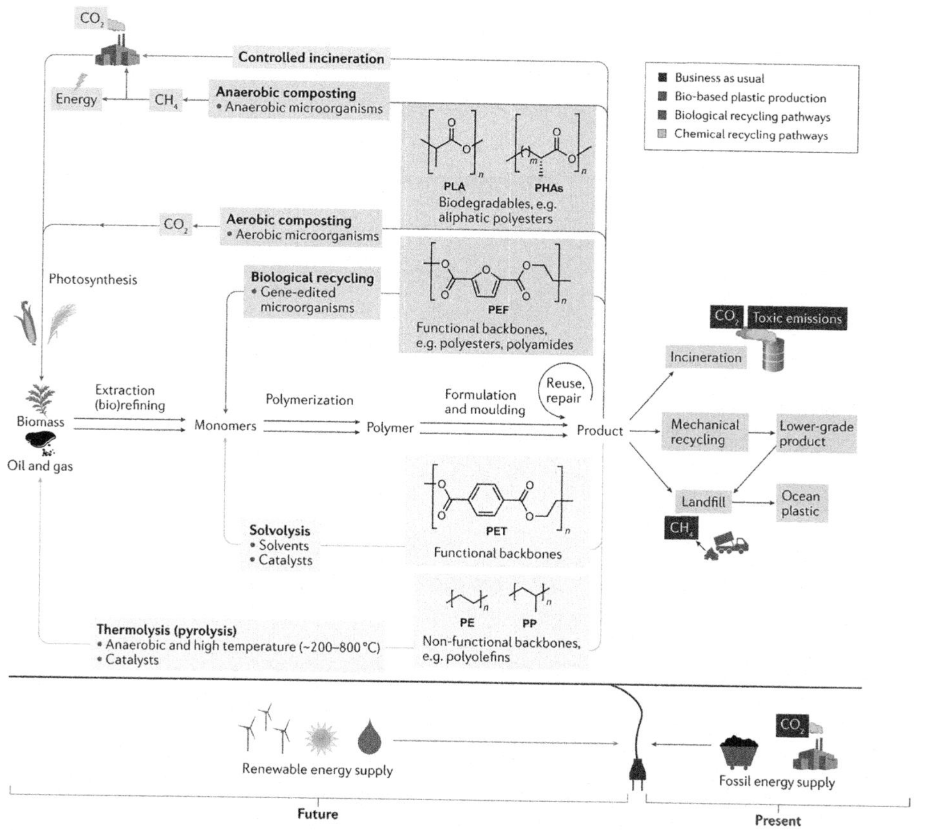

Fig. 8.5 The circular plastic economy. PLA: poly(lactic acid), PHA: polyhydroxyalkanoate, PEF: poly(ethylene furanoate), PET: poly(ethylene terephthalate), PE: polyethylene, PP: polypropylene. *Adapted from* [3] *with permission from Springer Nature*

guidelines for bioplastics, such as EN 16,760 (provides specific LCA requirements and guidelines for bio-based products, excluding food, feed and energy which is based on EN ISO 14040 and EN ISO 14044) [3]. A typical framework for LCA is shown in Fig. 8.6. Non-renewable energy use (NREU; unit: MJ/ kg polymer pellet/resin) and global warming potential (GWP; unit: kg CO_2 eq./kg polymer pellet/resin) have been commonly used for LCAs [22, 23]. Studies show that LCAs based on lab-scale data do not necessarily represent a future industrial-scale production; hence, ex-ante LCA can be used. Ex-ante LCA are studies that upscale an emerging technology (at a low technology readiness level) and compared them to an established one (Fig. 8.7) [24].

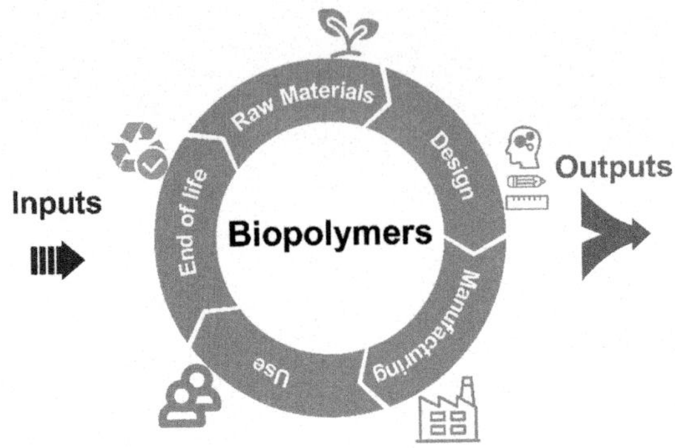

Fig. 8.6 Framework for life cycle assessment

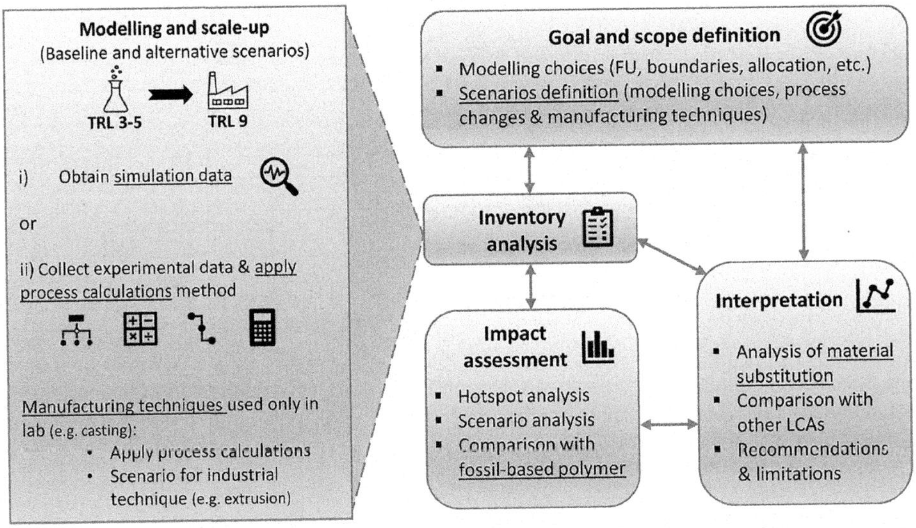

Fig. 8.7 Framework for ex-ante LCA of novel biopolymers at low TRL. LCA: life cycle assessment, FU: functional unit, TRL: technology readiness level. *Adapted from [24] with permission from Elsevier*

References

1. Biopolymers Market Forecast is Sunny with Explosive Growth. Plastics Today. Available from: https://www.plasticstoday.com/biopolymers/biopolymers-market-forecast-is-sunny-with-explosive-growth
2. Skoczinski P, et al. Bio-based Building Blocks and Polymers – Global Capacities, Production and Trends 2022–2027. nova-Institut GmbH (Ed.), Hürth, Germany2023.

3. Rosenboom J-G, Langer R, Traverso G. Bioplastics for a circular economy. Nature Reviews Materials. 2022;7(2):117–137.
4. Endres H-J, Siebert-Raths A. Engineering Property Profiles of Biopolymers. In: Endres H-J, Siebert-Raths A, editors. Engineering Biopolymers. Hanser; 2011. p. 149–224.
5. Endres H-J, Siebert-Raths A, editors. Engineering Biopolymers. Hanser; 2011.
6. Endres H-J, Siebert-Raths A. End-of-Life Options for Biopolymers. In: Endres H-J, Siebert-Raths A, editors. Engineering Biopolymers. Hanser; 2011. p. 225–243.
7. Thompson RC, Courtene-Jones W, Boucher J, et al. Twenty years of microplastics pollution research—what have we learned? Science. 2024;386(6720):eadl2746.
8. Vethaak AD, Legler J. Microplastics and human health. Science. 2021;371(6530):672–674.
9. Rochman CM. Microplastics research—from sink to source. Science. 2018;360(6384):28–29.
10. Microplastics are everywhere — we need to understand how they affect human health. Nature Medicine. 2024;30(4):913–913.
11. Consumer Guide to Recycling Codes. U.S. Depatment of Energy Office of Energy Efficiency & Revewable Energy 2021.
12. Samir A, Ashour FH, Hakim AAA, et al. Recent advances in biodegradable polymers for sustainable applications. npj Materials Degradation. 2022;6(1):68.
13. BPI. Biodegradable vs Compostable. Available from: https://bpiworld.org/biodegradable-vs-compostable
14. Gounder S. What Are The Different Certifications For Compostable Plastic? ETSUS. Available from: https://etsus.co/what-are-the-different-certifications-for-compostable-plastic/
15. BPI. Value of Certification. Available from: https://bpiworld.org/why-bpi
16. Australasian Bioplastics Association. Commericial Compostable Verification Programme. Available from: https://bioplastics.org.au/certification/the-seedling-logo/
17. Bioplastics - Industry standards & labels. European Bioplastics. 2023.
18. Japan BioPlastics Association. Certification Systems. Available from: http://www.jbpaweb.net/english/e-gp/index.html#aim02
19. TÜV Rheinland. Biodegradability Testing. Available from: https://www.tuv.com/world/en/biodegradability-testing.html
20. United Nations Industrial Development Organisation. Addressing the challenge of marine plastic litter using circular economy methods: relevant considerations. Available from: https://unido-gc.org/Publications/Addressing%20the%20challenge%20of%20Marine%20Plastic%20Litter%20using%20Circular%20Economy%20methods.pdf.
21. Curran M. Life Cycle Assessment: A Systems Approach to Environmental Management and Sustainability. CEP. 2015:26–35.
22. Yates MR, Barlow CY. Life cycle assessments of biodegradable, commercial biopolymers—A critical review. Resources, Conservation and Recycling. 2013;78:54–66.
23. Kookos IK, Koutinas A, Vlysidis A. Life cycle assessment of bioprocessing schemes for poly(3-hydroxybutyrate) production using soybean oil and sucrose as carbon sources. Resources, Conservation and Recycling. 2019;141:317–328.
24. Müller-Carneiro J, Figueirêdo MCBd, Rodrigues C, et al. Ex-ante life cycle assessment framework and application to a nano-reinforced biopolymer film based on mango kernel. Resources, Conservation and Recycling. 2023;188:106637.

Appendix A

Common functional groups

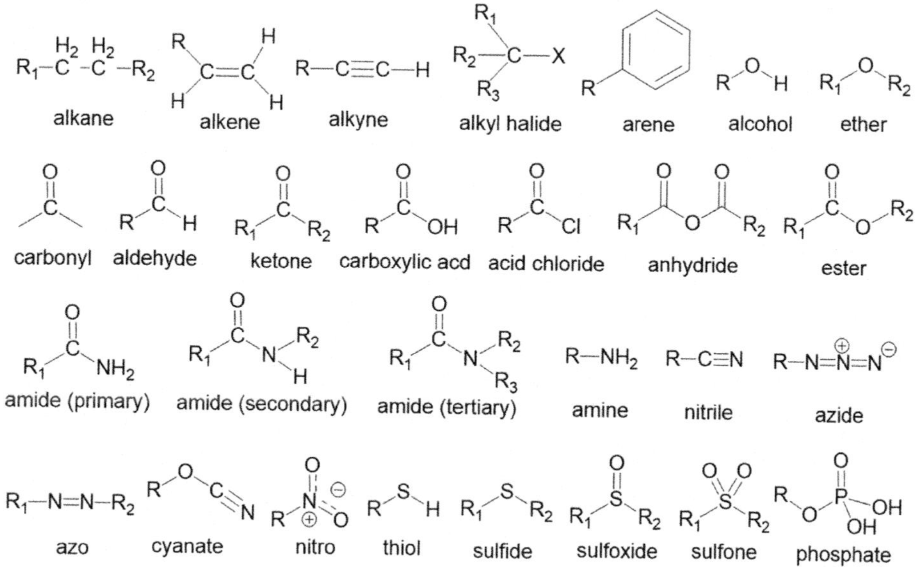

W. Zhang, *Biopolymer Engineering*, Synthesis Lectures on Chemical Engineering and Biochemical Engineering, https://doi.org/10.1007/978-3-031-98868-4

Appendix B

Summary of the equations for the step-growth polymerization

A-B system	A-A, B-B system
$M_0 = M_{r.u}$	$M_0 = \frac{M_{r.u.}}{2} = \frac{M_{01}+M_{02}}{2} = \frac{M_1+M_2}{2} - M_{s.m.}$
$X_n = DP$	$X_n = 2DP$
$X_n = \frac{1}{1-p}$ $(\uparrow p \rightarrow \uparrow X_n)$	$X_n = \frac{1+r}{1+r-2rp}$ $(\uparrow p \rightarrow \uparrow X_n; \uparrow r \rightarrow \uparrow X_n)$
$p = \frac{N_0-N}{N_0}$	$r = \frac{N_{Ao}}{N_{Bo}}; r \leq 1$
$< M_n > = X_n \times M_0$	
$< M_n > = \frac{1}{1-p} \times M_0$	$< M_n > = \frac{1+r}{1+r-2rp} \times M_0$

Note

DP: degree of polymerization (based on repeating units)

$M_{0\#}$: molar mass of monomer unit # (from monomer #)

$M_{\#}$: molar mass of monomer #

$M_{s.m.}$: molar mass of the small molecule produced (e.g., H_2O and HCl) during the polymerization

N: number of molecules remaining after certain time

N_0: Initial number of molecules present in the reaction system

p: fraction of monomer conversion; extent of reaction

r: ratio of the number of molecules of the reactants; stoichiometric imbalance

X_n: number-average chain length (based on structural or monomeric units)

W. Zhang, *Biopolymer Engineering*, Synthesis Lectures on Chemical Engineering and Biochemical Engineering, https://doi.org/10.1007/978-3-031-98868-4

Appendix C

The pk_a values and structures of amino acids

Amino acid	$pk_{a,1}$ (α–COOH)	$pk_{a,2}$ (α–NH_2)	$pk_{a,R}$ (side group)
Alanine (Ala/A)	2.34	9.69	
Arginine (Arg/R)	2.17	9.04	12.48
Asparagine (Asn/N)	2.02	8.80	
Aspartic acid (Asp/D)	2.09	9.82	3.86
Cysteine (Cys/C)	1.71	10.78	8.33
Glutamic acid (Glu/E)	2.19	9.67	4.25
Glutamine (Gln/Q)	2.17	9.13	
Glycine (Gly/G)	2.34	9.60	
Histidine (His/H)	1.82	9.17	6.00
Isoleucine (Ile/I)	2.36	9.68	
Leucine (Leu/L)	2.36	9.60	
Lysine (Lys/K)	2.18	8.95	10.53
Methionine (Met/M)	2.28	9.21	
Phenylalanine (Phe/F)	1.83	9.13	
Proline (Pro/P)	1.99	10.60	
Serine (Ser/S)	2.21	9.15	
Threonine (Thr/T)	2.63	9.10	
Tryptophan (Trp/W)	2.38	9.39	
Tyrosine (Tyr/Y)	2.20	9.11	10.07
Valine (Val/V)	2.32	9.62	

W. Zhang, *Biopolymer Engineering*, Synthesis Lectures on Chemical Engineering and Biochemical Engineering, https://doi.org/10.1007/978-3-031-98868-4

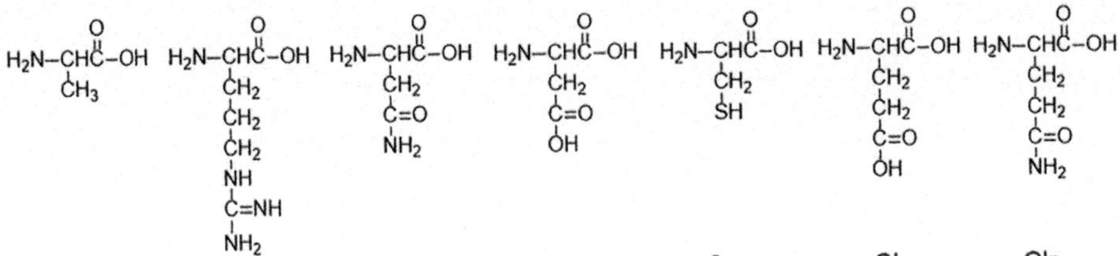

Ala Arg Asn Asp Cys Glu Gln

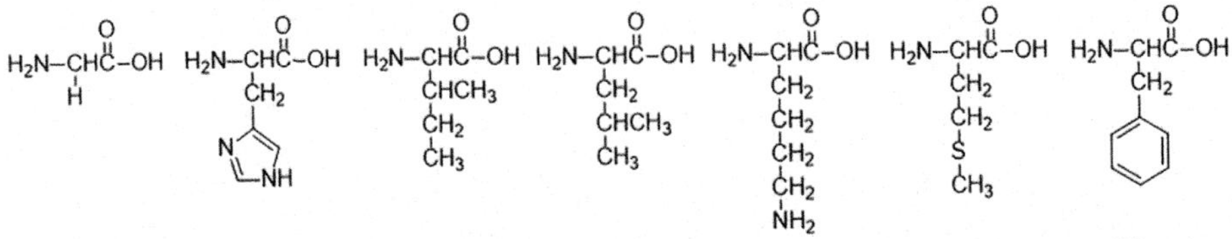

Gly His Ile Leu Lys Met Phe

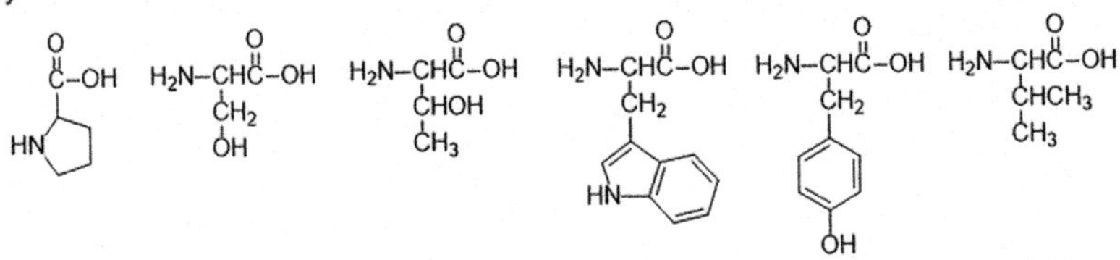

Pro Ser Thr Trp Tyr Val

Appendix D

Common chemical shifts in ^1H NMR spectroscopy

Hydrogen type	Chemical shift (δ), ppm
RCH_3	~0.9
R_2CH_2	1.2–1.4
R_3CH	1.4–1.7
RNH_2	1–5
$ArCH_3$	2.2–2.4
$R_2C{=}CH_2$	4.6–5.0
$R_2C{=}CHR$	5.2–5.7
RC≡CH	2–3
HC–OR	3.3–4
RCOO–CH	3.7–4.1
HC–COOR	2–2.2
H–C–C=O	2–2.7
ROH	1–5
RCOOH	10.5–12
$R_2C{=}CHR$	3.7–6.5
ArH	6.0–8.7
ArCH	2.2–3.0
ArOH	4–12
RNH_2	1–5

W. Zhang, *Biopolymer Engineering*, Synthesis Lectures on Chemical Engineering and Biochemical Engineering, https://doi.org/10.1007/978-3-031-98868-4

Appendix E

Common FT-IR bands and related function group assignments

Wavenumber, cm^{-1}	Functional group
3350–3450	O–H stretching of carbohydrate, proteins, and polyphenols
3000–3100	C–H stretching of aromatic compounds
2800–3000	C–H stretching of aliphatic compounds
2260–2210	C≡N stretching
2260–2100	–C≡C– stretching
1760–1665	C=O stretching (general carbonyls)
1760–1690	C=O stretching (carboxylic acids)
1750–1720	C=O stretching (ester)
1745–1715	C=O stretching (ketone)
1740–1720	C=O stretching (aldehyde)
1680–1600	–C=C– stretching (alkene)
1700–1600	Amide I (C=O stretching)
1560–1500	Amide II (N–H bending, C–N stretching)
1200–1300	Amide III (C–N stretching, N–H bending)

W. Zhang, *Biopolymer Engineering*, Synthesis Lectures on Chemical Engineering and Biochemical Engineering, https://doi.org/10.1007/978-3-031-98868-4

The manufacturer's authorised representative in the EU is Springer Nature Customer Service Centre GmbH, Europaplatz 3, 69115 Heidelberg, Germany. If you have any concerns regarding our products, please contact ProductSafety@springernature.com

Printed and bound by CPI Group (UK) Ltd, Croydon, CR0 4YY
31/07/2026
02182931-0001